Chipkin
Automation Systems

CW01307065

BACnet is a registered trademark of ASHRAE

LonWorks is a registered trademark of Echelon Corporation

Revision 1.0

Any reproduction or re-transmission in whole or in part of this work is expressly prohibited without the prior consent of Chipkin Automation Systems Inc

Copyright Notice

© Copyright 2009 Peter Chipkin who has given permission to Chipkin Automation Systems to publish this work.

Mailing Addr: 3495 Cambie St, # 211, Vancouver, BC, Canada, V5Z 4R3

Thanks to Tim Plumridge our cartoonist.

Chipkin
Automation Systems

Table of Contents

BACnet Objects and Properties	4
BACNET BBMD connects networks	8
Segmentation	9
BACnet Services	11
BIBB's	13
COV	17
MSTP	27
MSTP Bandwidth Issues	33
What can go wrong with RS485	37
MSTP Discovery	41
MSTP Slaves vs. Masters	43
Changing the Present Value	45
Trouble Shooting BACnet IP	49
Trouble Shooting BACnet MSTP	65
Hubs vs Switches	73
Resistors	75

Chipkin
Automation Systems

BACnet Introduction

Introduction
BACnet is an building automation and control networking protocol. It was developed by ASHRAE. BACnet was designed specifically to meet the communication needs of building automation and control systems. Typical applications include: heating, ventilating, and air-conditioning control, lighting control, access control, and fire detection systems.

History and Background
Who cares. It works. Its open and its growing

Flavors Of BACnet

Clarification

Native BACnet – What does that mean? In the Building Automation world the term 'Native' means that a device supports that functionality / protocol without the addition of a gateway or other module.

Flavor	Application	Affects you ?
IP	• Uses the IP protocol • Controller to controller • Controller to HMI • Some field devices.	On the up
Ethernet 802.3	• Raw Ethernet Packets	Being displaced by IP
Point to Point	• Modems and phone lines	Rare. Expect to disappear
MS/TP	• Field Devices	Millions of installed devices.
ARCnet	• Controller to controller	Rare. Expect to disappear

Chipkin
Automation Systems

BACnet Objects and Properties

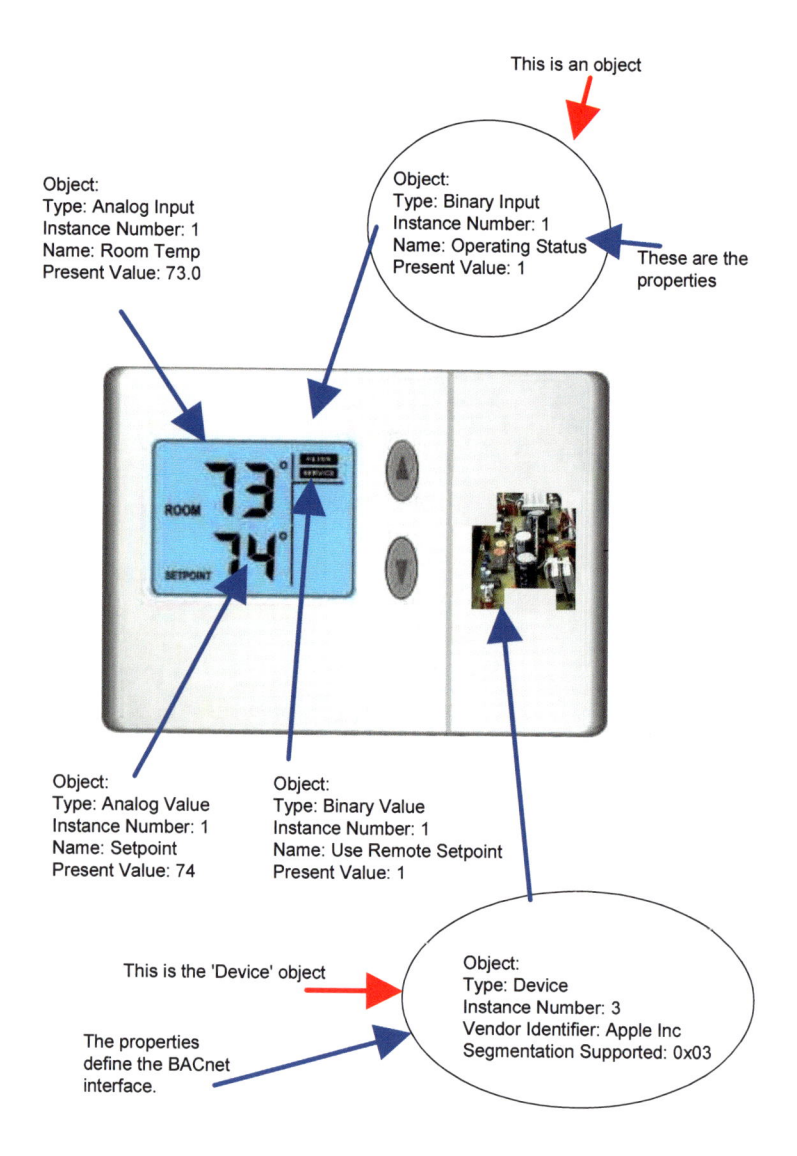

This is an object

Object:
Type: Analog Input
Instance Number: 1
Name: Room Temp
Present Value: 73.0

Object:
Type: Binary Input
Instance Number: 1
Name: Operating Status
Present Value: 1

These are the properties

Object:
Type: Analog Value
Instance Number: 1
Name: Setpoint
Present Value: 74

Object:
Type: Binary Value
Instance Number: 1
Name: Use Remote Setpoint
Present Value: 1

This is the 'Device' object

The properties define the BACnet interface.

Object:
Type: Device
Instance Number: 3
Vendor Identifier: Apple Inc
Segmentation Supported: 0x03

Page 4

Chipkin
Automation Systems

BACnet Objects and Properties cont'd

Data inside a BACnet device is organized as a series of objects. Each object has a type and a set of properties. There is always at least one object in a device – it is used to represent the device itself. The other objects represent the device's data.

In practical terms think of a simple thermostat. Our example is a simple device that has a temperature sensor, allows the set point to be changed locally or remotely, has a local remote selection and reports there is an internal fault by reporting its status as normal/abnormal.

Useful Tip

The **device object** is the first object read after a device is discovered because it has lots of interesting information for the client. For example, the device object has properties that report whether the device supports COV, whether more than one property can be read in a single message

Useful Tip

Unique Numbers are required for BACnet Device Object Instance Numbers across the entire network but in practice you can use duplicates as long as a routers separates then network segments the duplicates are connected to.

Heads Up

Some vendor systems and controllers require each object within a device to have a **unique name**. This is because they use the name as an internal index key (We agree that's stupid)

Chipkin
Automation Systems

BACnet Objects and Properties cont'd

Commonly used properties - Almost all objects you encounter will have these (and more) properties.

Object Type:
: Popular Object Types: Analog Input, Analog Output, Binary Input, Binary Output.

Instance Number:
: A number that must be not be repeated for any other object of the same type. The instance number and the object type must be unique for every object in a device.

Name:
: Speaks for itself.

Present Value:
: The current value of the object. BACnet has ways of telling you if the present value is valid – it uses a property called 'Reliability'.

Example of the properties and their values for a BACnet data object.

object-identifier:	analog-input [180]
object-name:	One_sec_Ph_A-NVolt
object-type:	analog-input
present-value:	100.000000
status-flags:	In-Alarm=[false], Fault=[true], Overriden=[false], Out-Of-Service=[false]
event-state:	normal
reliability:	unreliable-other
out-of-service:	False
units:	Volts
description:	Zero length/empty string

Chipkin
Automation Systems

BACnet Supports Discovery

In Modbus you need a data sheet to know what data is inside a field device. In BACnet you don't. You can go-online and discover the devices on a network and then interrogate the devices so they report what data objects they contain and what properties each object supports and what the current state of each property is.

The ability to discover is an almost universal truth in BACnet but there an obscure technicality may limit what you can learn about the object properties – Devices that support Read-Property-Multiple and don't support segmentation may not be able to fill in the response into a single message and thus don't respond with useful information.

We should also mention that most 'Discoverers' (or clients as we like to call them) cannot discover Vendor created proprietary properties.

There are two important practical implications of discovery

1. If your client software is half decent you do not have to type object/properties into the configuration screens. You simply discover and then drag and drop. Unfortunately more than half the BACnet software out there is not half decent.

2. You would think you can get away without data sheets but again you are then dependent on how decent a job the device Vendor did in naming and describing their points. Bad naming, missing object descriptions, un-implemented properties like max and min values make your job harder and force you to use Vendor docs.

Useful Tip

To discover devices on a **foreign subnet** you can configure the router to forward broadcasts or you can use BBMD

Chipkin
Automation Systems

BACNET BBMD connects networks

The BACnet discovery uses two services - called 'Who-Is' and 'I-Am'. They rely on the use of broadcasts.

Heads Up

Routers join IP networks together so messages from one network can be sent to another. Most routers do not forward broadcast messages and this means discovery **cant discover devices on another network.**

Routers join IP networks together so messages from one network can be sent to another. Most routers do not forward broadcast messages and this means discovery cant discover devices on another network.

To solve this problem BACnet provides a technology called BBMD - BACnet/IP Broadcast Management Device.

In overview the technology is simple. You install a BBMD (might be a physical device or just a software application on a computer) on each network. You can configure the BBMD by specifying the IP Address and mask of the each BBMD. This makes both BBMD configs identical. When the one BBM receives a broadcast, it forwards the messages to the other BBMD which in re-broadcasts on the other network. They are configured by BDT files and these may be modified on the fly using selected Bacnet services.

The technology also provides for foreign device registration. This allows a device on one network to communicate with a device on another network by using the BBMD to forward and route the messages.

Chipkin
Automation Systems

Segmentation in BACnet

Segmentation

BACnet messages that don't fit in a single packet use segmentation. Why would one message need more than one packet ? Well IP packets have a maximum length of 1500 bytes. So if you are sending a BACnet IP message that is longer than 1500 bytes then you need to send more than one Ethernet packet.

Example: Ask your fishing buddy if he wants another beer. The reply is short and fits in a single response packet. Yes. Now ask him to tell you about the one that got away. He will need multiple sentences to tell you the long story. Your buddy needs you to support segmentation otherwise you will only hear the first sentence of his storey (lucky you.)

Be Aware

1500 is not a hard coded Maximum Transmission Unit (MTU) length in all Ethernet applications. Often the size is set smaller.

Most serial protocols like MS/TP choose a small number for the MTU because an error requires retransmission and the data is slow so it better to catch an error in smaller packet.

How does Segmentation affect you as a user ?

If a device has a large number of objects and a message is sent to read the object list then it is possible that the response wont fit in a single packet. If both the device and the requestor support segmentation then there is no problem. If either side doesn't support segmentation then 1) You are out of luck or 2) The requestor must use a different method to read the object list - for example, reading each object using its index until it reaches an index number with no object.

The CAS BACnet explorer works like that - first it tries the most efficient method and then it slowly downgrades itself to try and ensure the response will fit in a single packet. Not all software works in this intelligent manner.

Chipkin
Automation Systems

Segmentation in BACnet cont'd

How do you know if a device supports segmentation ?

You can read the vendor's PIC (Protocol Implementation and Conformance) Statement or you can look at the device object's properties.

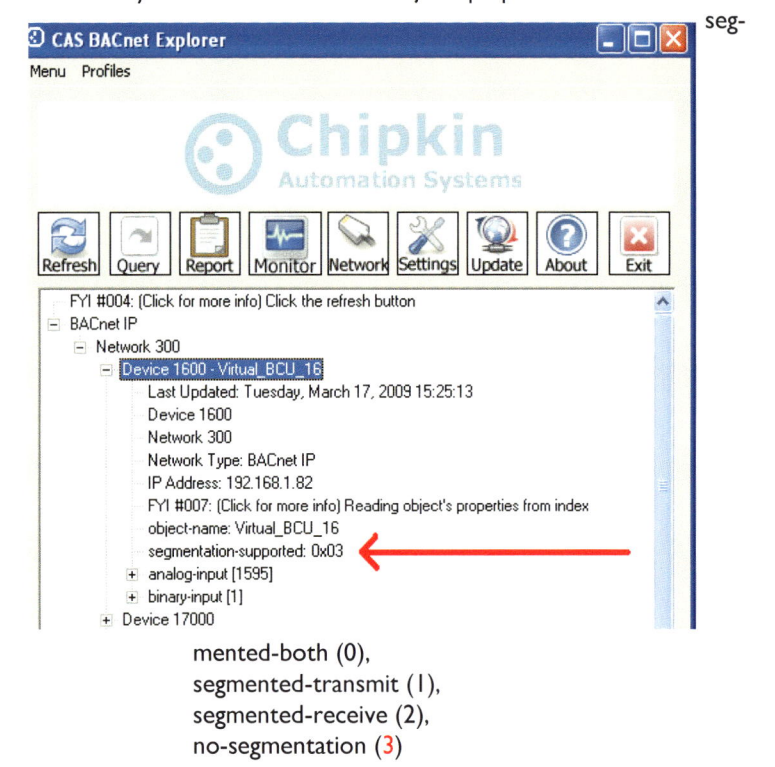

seg-mented-both (0),
segmented-transmit (1),
segmented-receive (2),
no-segmentation (3)

How can you work around the segmentation issue ?

Ensure you use read-property for a single property - avoiding read-property for all properties. If you have to use read-property-multiple then limit the list of properties to be read and avoid reading all using this service. This might not be possible if you cant configure which services get used by your system.

Page 10

Chipkin
Automation Systems

BACnet Services

Think of a service as a task / action. Reading and Writing data uses some of the BACnet services known collectively as Data Access Service.

BAS systems interact with BACnet devices and objects using these services and for the most part the actual service used is hidden from the engineer building / using the BAS.

When would you take an interest in what services are supported ? From our point of view (as user's of BACnet) now we can make the equation Services == Capabilities.

Example: Bacnet provides a service called 'CreateObject'. Most field devices do not support this service. Imagine that a BAS vendor provides an option to 'Create a new object in Field Device'. You try it and it produces an error since the device doesn't support the service. If you are fortunate the BAS you are working with will provide an error message that tells you why there was an error.

This is the point where the services supported by the BAS and the field devices becomes interesting to you. The place to look to see what services a device supports is in the vendor PIC statement. There is only one mandatory service - Read Property Service.

Unfortunately because the actual services used by a BAS are hidden beneath the GUI it's not always easy to know if you can exploit the capability of a field device.

Chipkin
Automation Systems

BACnet Services cont'd

Useful Tip

ReadPropertyMultiple - Reads one or more properties from one or more objects BUT there is a catch. If segmentation isn't supported then the reply must fit in a single transmission unit (typical effective for Ethernet is < 1500 bytes and for MSTP < 245 bytes.) This is a fundamental weakness in BACnet since you cant know in advance how big the response will be. A number of (cleverer) clients will use ReadProperty to read each item when a ReadPropertyMultiple fails

Service	Notes
AddListElement Service	Rarely supported
RemoveListElement Service	Rarely supported
CreateObject Service	Rarely supported
DeleteObject Service	Rarely supported
ReadProperty Service	Always Supported
ReadPropertyConditional Service	Rarely supported - Returns a list of objects/properties that meet the specified selection criteria. Eg all Analog Inputs.
ReadPropertyMultiple Service	Often supported
ReadRange Service	Rarely supported
WriteProperty Service	Almost always supported
WritePropertyMultiple Service	Less often supported than ReadPropertyMultiple

Chipkin
Automation Systems

BIBB's

The implementation of BACnet is under-the-hood. It is invisible to most of us users. We don't know what services are supported and nor do we know when each service is used. If we don't know then how can we tell, for example, whether a controller and field device such as a thermostat can interact.

The question is: How do you match requirements of a project to the capabilities of the devices being installed ? The Answer - in BACnet - is BIBBS.

A BIBB is a Bacnet Interoperability Building Block.

Continuing with our example: Say your controller needs to read the set point on a thermostat to perform its control. Then the controller needs to support a BIBB called DS-RP-A. This isn't enough. The thermostat must be able to respond so it needs to support a BIBB called DS-RP-B. DS stands for Data Sharing. RP stands for Read Property. The A and the B stand for Client (A) and Server (B).

The outdoor temperature on a HVAC controller could come from two places. A local sensor connected to the unit or a remote value sent via BACnet. In this case you would want to match the BAS controller and HVAC controller with BIBBS DS-WP-A and DS-WP-B (DataSharing-WriteProperty-A or B for Client and Server).

Thus, if you buy a great controller that supports Conditional and Range Reads you would want to buy field devices that support these services. DS-RPC-A/B for DataSharing-ReadPropertyConditional-Client/Server.

Abbreviation	BIBB Category
DS	Data Sharing
AE	Alarm and Event Management
DM	Device Management DM and NM are part of the same category.
NM	Network Management
T	Trend
SCHED	Schedule

BIBB's continued

Useful Tip

Device A and Device B. All the BIBBs end in '-A' or '-B'. BACnet refers to the A device and the B device. The A device in this context is a device acting as a client. The B device means a device acting as a server. In BACnet devices can be both clients and servers since almost all devices can read data from each other.

Extract – The full BIBBS table is available at http://www.chipkin.com/articles/bacnet-bibbs-table-bacnet-interoperability-building-blocks

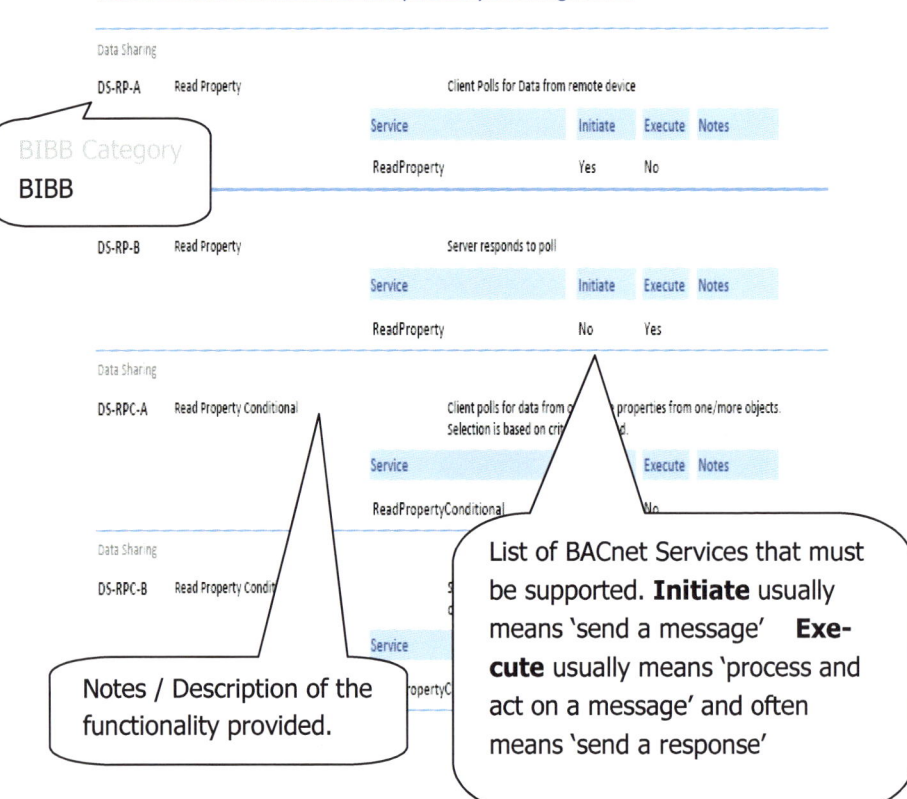

Data Sharing

DS-RP-A   Read Property   Client Polls for Data from remote device

Service	Initiate	Execute	Notes
ReadProperty	Yes	No	

BIBB Category: **BIBB**

DS-RP-B   Read Property   Server responds to poll

Service	Initiate	Execute	Notes
ReadProperty	No	Yes	

Data Sharing

DS-RPC-A   Read Property Conditional   Client polls for data from ... properties from one/more objects. Selection is based on cri...

Service	Execute	Notes
ReadPropertyConditional		

Data Sharing

DS-RPC-B   Read Property Condit...

Service

Notes / Description of the functionality provided.

List of BACnet Services that must be supported. **Initiate** usually means 'send a message' **Execute** usually means 'process and act on a message' and often means 'send a response'

Chipkin Automation Systems

PIC Statements

The BACnet specification defines theoretical capabilities. The PICS statement tells you what capabilities have been implemented.

Every BACnet capable device (Controllers, Software and field devices) has a PICS statement. The PICS statement is very useful to you as a Field Tech, Engineer or designer.

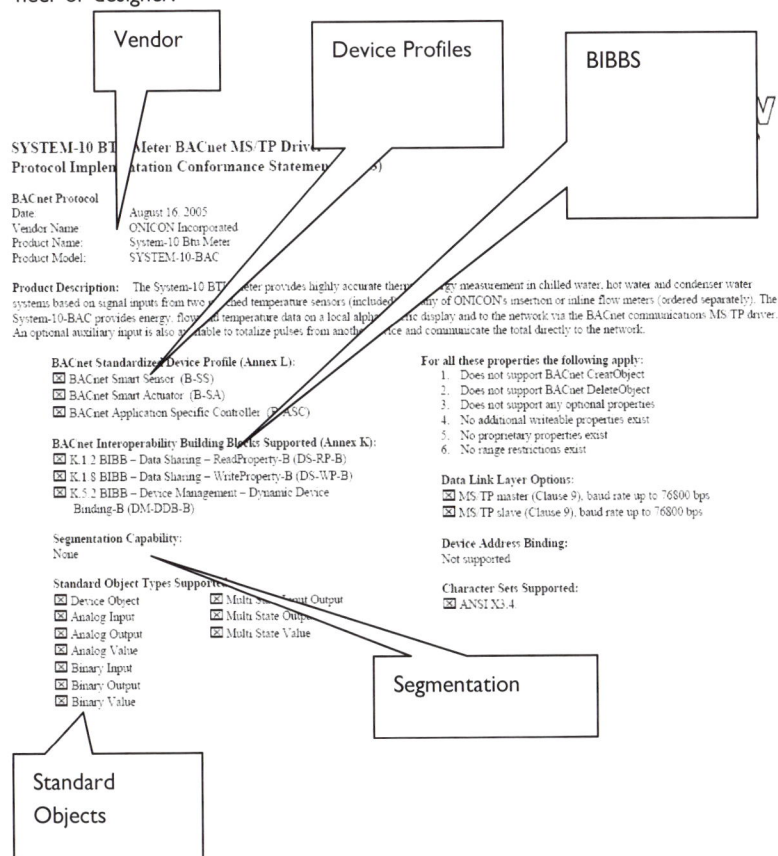

Vendor — **Device Profiles** — **BIBBS**

SYSTEM-10 BTU Meter BACnet MS/TP Driver
Protocol Implementation Conformance Statement (PICS)

BACnet Protocol
Date: August 16, 2005
Vendor Name: ONICON Incorporated
Product Name: System-10 Btu Meter
Product Model: SYSTEM-10-BAC

Product Description: The System-10 BTU Meter provides highly accurate thermal energy measurement in chilled water, hot water and condenser water systems based on signal inputs from two matched temperature sensors (included) and any of ONICON's insertion or inline flow meters (ordered separately). The System-10-BAC provides energy, flow and temperature data on a local alphanumeric display and to the network via the BACnet communications MS/TP driver. An optional auxiliary input is also available to totalize pulses from another device and communicate the total directly to the network.

BACnet Standardized Device Profile (Annex L):
☒ BACnet Smart Sensor (B-SS)
☒ BACnet Smart Actuator (B-SA)
☒ BACnet Application Specific Controller (B-ASC)

BACnet Interoperability Building Blocks Supported (Annex K):
☒ K.1.2 BIBB – Data Sharing – ReadProperty-B (DS-RP-B)
☒ K.1.8 BIBB – Data Sharing – WriteProperty-B (DS-WP-B)
☒ K.5.2 BIBB – Device Management – Dynamic Device Binding-B (DM-DDB-B)

Segmentation Capability:
None

Standard Object Types Supported:
☒ Device Object   ☒ Multi State Input   ☒ Multi State Output
☒ Analog Input   ☒ Multi State Value
☒ Analog Output
☒ Analog Value
☒ Binary Input
☒ Binary Output
☒ Binary Value

For all these properties the following apply:
1. Does not support BACnet CreateObject
2. Does not support BACnet DeleteObject
3. Does not support any optional properties
4. No additional writeable properties exist
5. No proprietary properties exist
6. No range restrictions exist

Data Link Layer Options:
☒ MS/TP master (Clause 9), baud rate up to 76800 bps
☒ MS/TP slave (Clause 9), baud rate up to 76800 bps

Device Address Binding:
Not supported

Character Sets Supported:
☒ ANSI X3.4

Segmentation

Standard Objects

Page 15

Blank Page

BACnet Change of Value (COV)

Introduction

Most field devices are passive servers. They wait passively for a system to poll them for data and only then do the devices respond. A consequence of this is that the data client only knows the value of an object property when it polls for the value. If the duration of an event (change of value) is shorter than the interval between polls then the data client will not know the event occurred.

BACnet provides a solution for this by defining services to report events. These services allow a device to be transformed from a passive server to an active server since it is now capable of sending messages reporting the value of an object property based on some event rules. BACnet COV is a sub-set of the 'Alarm and Event Services'.

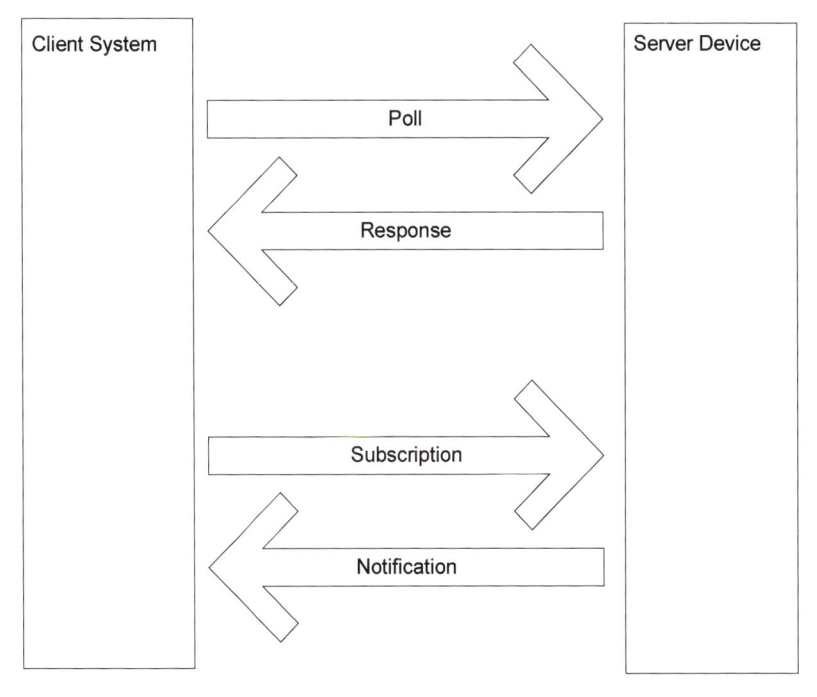

Chipkin
Automation Systems

This article discusses how the technology operates and then provides a discussion on some weaknesses in the COV system.

BACnet provides a solution for this by defining services to report events. These services allow a device to be transformed from a passive server to an active server since it is now capable of sending messages reporting the value of an object property based on some event rules. BACnet COV is a sub-set of the 'Alarm and Event Services'.

Simple Definition of COV

Data clients subscribe to an object for COV reporting. The device monitors the value of the object property, monitors the subscription list and the change criteria. When the change criteria are met the devices sends notifications of the new value to the subscribers.

Useful Tip

Not all devices / objects support COV
The BACnet COV system is not a mandatory part of the protocol. Each vendor decides if they want to support it. In addition, each vendor gets to decide which properties of which objects support COV.

The device object of a COV server indicates whether there is support for COV. Beyond that you have to look for the presence of certain

Page 18

How COV Works

The system is a little more complex than these note describes. For example, subscriptions have durations and there is more than one way the notification can be sent,

Subscribe
Data Client sends messages to the device to subscribe to COV notifications. The server must accept the subscription

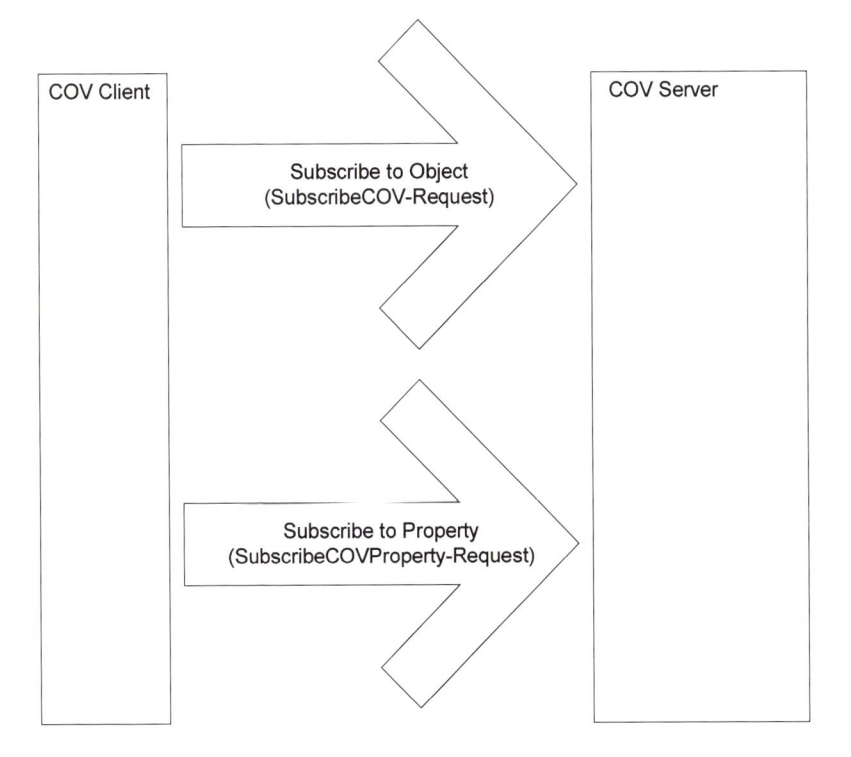

```
COV Client                                              COV Server

              Subscribe to Object
              (SubscribeCOV-Request)  >


              Subscribe to Property
              (SubscribeCOVProperty-Request)  >
```

Monitor
COV Server device monitors the property values of subscribed objects and the COV change criteria.

How COV Works cont'd

Notify
When a change has occurred that meets the COV change criteria the server sends notifications to the subscribers or for those objects with a COV Period defined, notifications are sent based on the period if there has been no change.

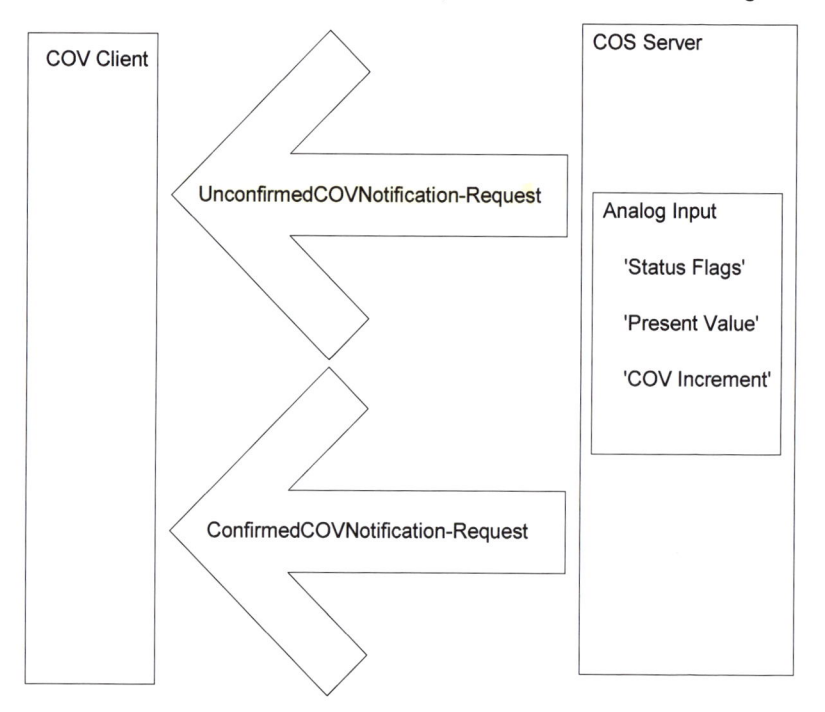

```
COV Client                                    COS Server

           <-- UnconfirmedCOVNotification-Request --

                                              Analog Input
                                              'Status Flags'
                                              'Present Value'
                                              'COV Increment'

           <-- ConfirmedCOVNotification-Request --
```

Process Notification
The data client must process the notification and update the display, log ....

Unsubscribe / Renew Subscription
If the data client no longer needs the subscription it unsubscribe. If it needs the subscription maintained then the client should periodically re-subscribe.

Chipkin
Automation Systems

Key Elements of the COV Technology

COV Server :
A BACnet device that supports COV, accepts subscriptions and sends COV notifications messages to a COV Client.

COV Client :
A BACnet device, typically a SCADA or Logging application, which can subscribe for COV notification and which can process the notification messages.

Subscription :
Establishes a relationship between a COV Server and a COV Client.
Subscriptions have the following attributes ;
    Subscription to an Object or to a Property
        BACnet provides two services for subscription. One subscribes to an object and the other to a property of an object.
    Identification of the Subscriber
        The server needs to know where to send the notifications.
    Object Identifier
        Eg. Analog Input 1
        If subscribing to a property then the property must be identified too.
    Lifetime
        Is indefinite or a specific number of seconds. Values can be large.
    Notification Type
        Notification can be sent with/without requiring confirmation from the data client.
    COV Increment
        This parameter is only used in subscriptions to object properties. If not specified in the subscription the object uses its own increment.

Notification :
A message which reports the current value of the changed property as well as the current state of the objects Status Flag property if it exists. The notification also contains the number of seconds remaining to the subscription. Confirmed notifications require a response from the COV client.

COV Change Criteria

Useful Tip

The device needs to send notification of a change to you. How does it decide when to do this ? You have some control—specify the COV Increment when you subscribe. You should also consider the object and property type because there are different rules for when they send notifications.

The change criteria are based on the type of subscription

Subscribe to the object :
    Either of these changes trigger notification

      1) If the status flags change at all
      2) If the Present Value changes

- Binary, Life Safty and Multistate Objects : any change to Present Value
- - Analog, Loop and Pulse Objects: change by COV_Increment

Subscribe to a property :
    Either of these changes trigger notification

      1) If the status flags change at all (if the object has status flags)
      2) If the Property Value changes

- Property is of type REAL : change by COV_Increment (which may be defined in the subscription or may be the native increment defined in the device).
- - Property is of some other type: any change to Present Value

COV Issues and Limitations

General Discussion

There are some intrinsic problems with event or change reporting systems.

In our experience it is fairly common to learn that a well configured system failed to deliver the critical information only after a significant failure. We learn the hard way that we polled for data too slowly, the logger was offline when the notification was sent or that the logger was swamped with data because of the failure of incoming feed. With this in mind we draw your attention to some if the issues you should consider.

The deluge of changes problem

These event reporting systems are commonly implemented to reduce the bandwidth requirements for monitoring remote devices or to ensure that the data client sees changes whose durations are less than their minimum polling interval. When a client reduces the frequency of its polls for data or reduces which objects it polls this reduces the bandwith requirements of the system. To compensate for the lower frequency of updates a COV system may be employed.

If COV systems are employed to 'guarantee the delivery of critical data' great care should be taken is assessing the so called 'guarantee'. The guarantee is often based on the assumption that changes are infrequent and small in scope but this isn't always the case. Often a single change can spawn a number of changes and those changes can spawn more changes in a system similar to a nuclear reaction. The more changes that occur the more notifications that must be sent. If all the changes occur in a short interval then it easy to foresee a situation where a network or data client can be deluged with notifications. Now we have to assume there is enough bandwidth to handle sending them all and that each client can handle all the incoming messages quickly. If the notifications require confirmation then the speed of the client in processing the messages is material. If there is no notification required then it is conceivable in a poor BACnet implementation that the client could drop messages when its buffer is full.

It is very difficult to test these conditions because the test requires monitoring large data sets and the test setup requires a knowledge of how changes can spawn other changes within a device and / or a system. A consequence of these difficulties is that the performance of COV systems can be unpredictable.

COV Issues and Limitations Cont'd

The offline Subscriber problem

A subscriber may be offline when a notification is sent. If the subscription required confirmed notification then this could present a temporary but significant loss of bandwidth while the COS server waits the timeout period before sending the next notification. If no confirmation was required then there is no way of knowing that the subscriber was offline and even if it wanted to the COS server device cannot signal this in anyway.

The COV system does not require vendors to manage notifications sent to offline subscribers. Thus the COV server does not have to queue them and resend them. Thus, it is possible for a COV client to lose its data synchronization with the Server. The only way around this is to use occasional polling. However, if the data client was a logging system then the damage is already done and the log records will not be made.

Useful Tip

The device needs to send notification of a change to you. How does it decide when to do this ? You have some control—specify the COV Increment when you subscribe. You should also consider the object and property type because there are different rules for when they send notifications.

Page 24

BACnet COV Limitations

Subscriptions may be lost on reset

The BACnet spec does not require vendors to maintain the subscriptions if the device resets. Thus a reset may result in the loss of all subscriptions. Now the system is dependent on the frequency of the re-subscription by the data client. That frequency is a vendor choice too and some systems don't send periodic re-subscriptions.

Single/Multiple Subscriptions

A single data client can potentially subscribe more than once to the same data object. This is possible because in identifying itself, in the subscription, the client provides two elements of information - identifier and handle. The identifier tells the server where to send responses. The handle is a number allocated by the client for some internal purpose. Changing the handle is enough to make a subscription unique and hence it is possible to have multiple subscriptions to the same object from the same client.

Subscribing to Arrays

The spec allows vendors to choose whether to support subscriptions to all or particular elements of the array.

Variable Number of Subscriptions

Each vendor chooses how many subscriptions to an object / property are supported. The spec requires that at least one must be supported. You should assume the list is finite and fairly short.

Its conceivable that subscription space could be wasted by temporary subscriptions from test equipment of software thus denying room for important subscriptions.

It is important to understand how your COV client application handles and reports failed subscriptions as these events can be as important as the event they are attempting to monitor.

COV Issues and Limitations Cont'd

Unconfirmed Notifications can be sent without Subscription

The spec allows a device to send unconfirmed notifications for any property of any object to any other device on the network. Thus a device can send changes of a object property that has a wide area interest such as an outside air temperature. The vendor chooses if they wish to implement this, which objects, which properties and the frequency of the notification.

Notifications can be sent without a change of value

The criteria for triggering notification messages requires notifications to be sent if the Status Flags associated with the object change even if the value hasn't changed. This is potentially, important information but a number of data client systems ignore this element of the notification data.

Warning !

There are some risks to using COV. Read the notes on these pages to learn about them. We strongly recommend that you test the system to avoid being burned by assumptions.

If data is critical then poll for it as well as using COV.

Chipkin
Automation Systems

MSTP— An Introduction

This flavor of BAcnet is most commonly used to connect field devices to controllers / routers / control applications.

MSTP

M = Master
S = Slave
TP = Token Passing

The physical layer uses RS485 which allows up to 128 devices to be installed on a single network with a max physical length of 4000ft and speeds up to 115k baud. Using repeaters allows the length to be increased. Compare to Ethernet where the spec allows a max of 100 meters (330ft) on a single unrepeated segment.

Common baud rates are 19200, 38400 and 76800. All devices must operate at the same baud rate. More and more devices can auto sense the baud rate and configure themselves correctly.

Warning !

A number of microprocessor UARTS cannot accurately produce 76800 baud signals. Devices using these microprocessors might list 76800 as supported. If you are having issues you might want to downgrade your network.

We divide the messages on a MSTP network into two categories
- Overhead (token, poll for master...)
- Application. These carry payloads that we have an interest in.

Only a device with the token can initiate an application layer message. It can send the message to any device on the network. Some messages demand an instantaneous reply, some don't. The receiving device doesn't need the token to respond. There is a limit to how many application layer messages a device can send before it must pass the token on. (The limit is implied by the max number of frames that can be sent before the token is passed.))

MSTP— An Introduction

The benefits of token passing networks are the following

1. They are self healing
2. They can discover new devices
3. They ensure each device gets its chance

They avoid collisions making network performance (somewhat) deterministic

A disadvantage of the token system is that any one device gets a limited use of the bandwidth. Thus a device may need to keep an internal queue of application layer messages it wants to send, waiting to use the token. We have encountered some vendor systems which fill there queue and then drop subsequent messages without notifying the user of the problem. Limited access combined with the overhead makes it easy to use up all the bandwidth on the network if there are many devices with many objects and many properties of interest.

⊙ Chipkin
Automation Systems

BACnet MSTP Installation and RS485

Here is our simplified advice :

RS485 is a 3 conductor network.
You take a huge risk by not installing the 3rd conductor. You risk blowing 485 ports, you risk unstable operation (works sometimes and doesn't work other times) and finally you risk re-installation. For a more detailed discussion read this article http://www.chipkin.com/articles/rs485-cables-why-you-need-3-wires-for-2-two-wire-rs485 . The more power sources used to power devices, the greater the physical separation of devices, the less well grounded devices and power sources are the greater the risk. Remember this statement : The so called Ground Terminal on a RS485 interface is not a connection to ground. It is a common reference signal. The voltage level on the Tx/Rx conductors are measured relative to this voltage level.

You can (if you must) use a shield drain wire as the 3rd conductor (ground reference conductor)

Controversial

Some people argue that the 3rd conductor is not necessary. We argue it is. Read more about this debate here:
http://www.chipkin.com/articles/rs485-cables-why-you-need-3-wires-for-2-two-wire-rs485

Ground Reference
Always connect the ground reference conductor first if you are connecting a device that is powered up or you are connecting your laptop an operating network.

OR

Always choose devices that have optical isolation - this almost always will protect the RS485 transmitter / receivers.

Cable Shield
You can get away without the shield. The twisted pair used for Tx and Rx is more effective at noise cancellation than the shield.

Chipkin
Automation Systems

MSTP— Where to run your cables

Cable Runs

Take care where you run your cables. It seems obvious not to wind your cable around other cables or sources of electricity / magnetism. People are often surprised to find that the worst source of induced noise are switching DC loads. Another big culprit are Variable Frequency drives

This is the RS485 Cable. It not secured so vibration will get a chance to do its work,

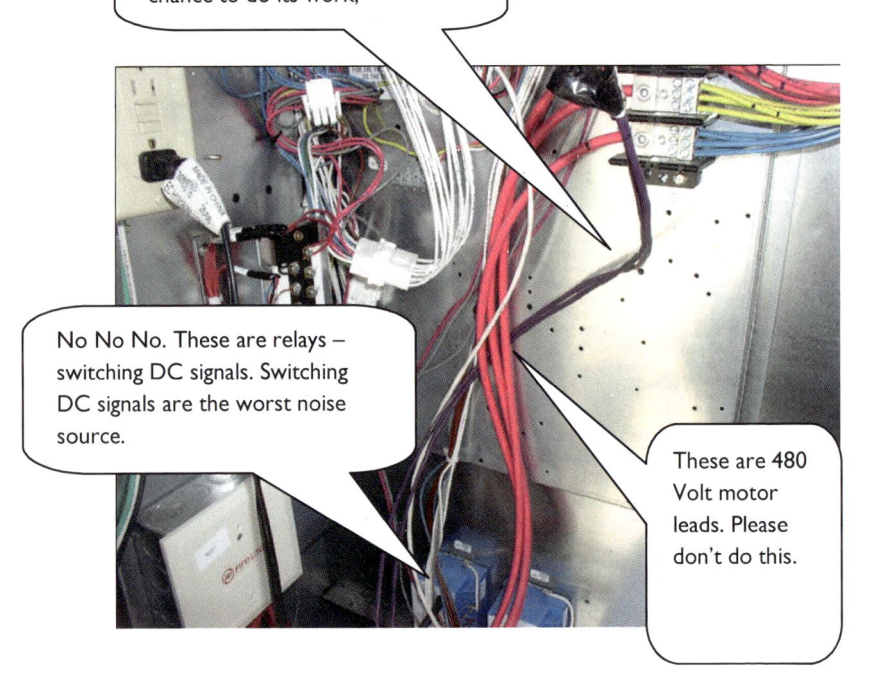

No No No. These are relays – switching DC signals. Switching DC signals are the worst noise source.

These are 480 Volt motor leads. Please don't do this.

Page 30

Chipkin
Automation Systems

BACnet MSTP Cable Selection

Cable selection does make a difference.

All cables offer impedance (resistance). Some cables are designed so that the impedance is relatively independent of distance. You want one of these cables. A clue to knowing if you selected one is to look at the cable's Nominal Impedance. If they quote a number such a 100Ohms you have a good cable. If they quote an impedance per meter/foot you have chosen the wrong kind. Wrong in the sense – to determine the value of terminating resistors now requires measurements and calculations. Choose low capacitance cables.

Can you use Cat5 cable ? Yes. Use one pair for Tx,Rx and a conductor from another pair for the ground reference signal.

We recommend these two cables.

Belden 3106A

Multi-Conductor - EIA Industrial RS-485 PLTC/CM
22 AWG stranded (7x30) tinned copper conductors, Datalene® insulation, twisted pairs, overall Beldfoil® shield (100% coverage) plus a tinned copper braid (90% coverage), drain wire, UV resistant PVC jacket.

Belden 3107A

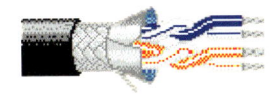

Multi-Conductor - EIA Industrial RS-485 PLTC/CM
22 AWG stranded (7x30) tinned copper conductors, Datalene® insulation, twisted pairs, overall Beldfoil® shield (100% coverage) plus a tinned copper braid (90% coverage), drain wire, UV resistant PVC jacket.

Chipkin Automation Systems

BACnet MSTP Cable Lengths

Cable Lengths and Baud Rates

Practically speaking you can go up to 4000 feet at baud rates up to 76800 baud. Above that you need to do a little math and reduce the length. For example, at 115k baud your cable should not be much longer than 2500 feet.

However, the higher the baud rate the more sensitive the cable is to the quality of installation - issues like how much twisted pair is unwound at each termination start to become very very important.

Our advise: For longer networks with lots of devices, choose 38k400 baud over 76k800 baud and optimize using COV, separate networks and by setting the Max Master to a lower number.

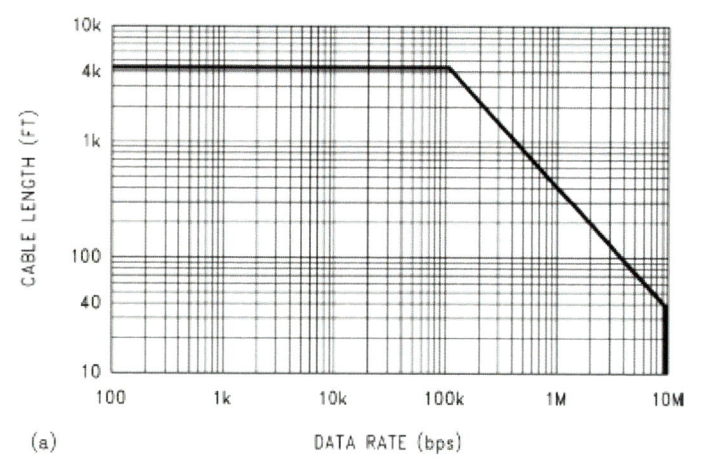

*Source: **Ten Ways to Bulletproof RS-485 Interfaces** National Semiconductor Application Note 1057 John Goldie October 1996*

BACnet MSTP Bandwidth Issues

There are non electrical considerations to determine how many devices you put on an MSTP network.

The chart below illustrates (from one installation) how little of the bandwidth is used to transfer data. The APDU's are application layer message that poll and respond with property values - they do work for us as data consumers. The rest is used to maintain the network - passing the token around and looking for new devices.

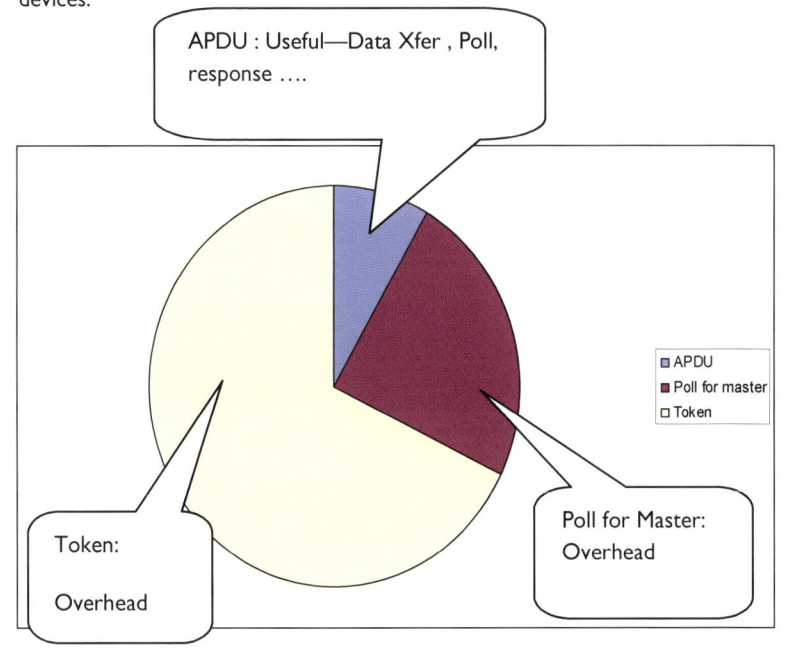

APDU : Useful—Data Xfer , Poll, response ....

- APDU
- Poll for master
- Token

Poll for Master: Overhead

Token: Overhead

It's not possible to provide a calculator to work out how many devices to install on a single network but the following list provides some help in assessing bandwidth considerations.

BACnet MSTP Bandwidth Issues cont'd

How many of the devices will be BACnet slaves.

Token passing and looking for new devices on the MSTP trunk consumes a fair amount of bandwidth.

A BACnet slave can be read/written but never gets the token. So it can't initiate any messages because it never gets the token. The more slaves the fewer token passes. Typically you are not able to put a device in slave mode. Most vendors implement their devices as masters (ie token passing devices)

How many Objects in each device are you interested in monitoring ?

The more you read and the greater the frequency the more bandwidth that will be consumed.

It takes approx 30 bytes to poll for a single property. It takes about 40 bytes to reply. A token is 8 bytes as is a Poll for master. Assume that 50% of your bandwidth will be used by overhead (token, poll for master). Divide the baud rate by 10 to get bytes per seconds. Using a number like 30+40=70 as a best case and 100 as a worst case (obviously reading descriptions will take more) multiply by the number of objects and properties you are going to poll on a regular basis.

Baud	38400
Bytes Per Sec	3840
Overhead	50%
Bytes per Sec for Payload	1920
Typical poll and response for a single property	70
Number of properties that can be polled per sec	24.43
Typical number of properties that must be polled per object	4
Number of objects per sec	6.86

Chipkin
Automation Systems

BACnet MSTP Bandwidth Issues cont'd

How many properties from each of these objects ?

What is the baud rate?

What is Max Master Set to?

> Every few cycles each (master) token passing device on the network must look to see if there are new devices. Max Master determines the biggest address that must be searched for. Each search involves sending a message and waiting for a response or a timeout (if the devices isn't there). Timeouts cost time. The higher the number of Max Master (default is 127) the more potential devices must be searched for. If you use Max master to improve bandwidth then you must adjust it in each device.

Warning !

> Every device can have max master set to a different value.
>
> Max Master is your friend in saving some bandwidth but your enemy when it limits the discovery of new devices. No device with a MAC address greater than max master will be discovered.

Do the devices support the "Read/Write Property Multiple" services or must each property be read in a separate message.

> Find the answer to this question by reading the BIBs statement for each device or you could explore the device object of the device, find the property called BACnetServicesSupported and then look at the 14th item in the array to see if Read Property Multiple is supported and the 16th for Write Property Multiple. However, we have found that a large number of devices don't display this information.
>
> Obviously, if you can read a chunk of properties in one message you will be better off than if you can only read a single one.

Chipkin Automation Systems

BACnet MSTP Bandwidth Issues cont'd

Can you use BACnet's COV mechanism.

> COV stands for Change of Value. When a device supports COV another device / application can subscribe to receive notifications when an object property changes. This means the data client doesn't have to poll for data continuously but can wait passively to be notified of the change. This reduces the number of messages on a network dramatically.

Some devices are slower than others.

> BACnet allows up to 15 msec for a device to use the token. Since most messages on a MSTP network are token passes a device that uses the token in 5 msec will consume much less bandwidth than one that takes 15msec. (A number of vendors relax this requirement to allow for other vendors implementations. The more relaxed the more bandwidth is consumed doing nothing.)

How do you put more than 32 devices on a single RS485 trunk ?

The simple answer is use a repeater but in practice one isn't always necessary.

Useful Tip

Use a RS485 repeater to put more than 32 devices on a trunk. The repeater doesn't need to know its repeating BACnet messages.

The RS485 standard is based on 32 devices. Since the standard was developed most RS485 chips present less than the full unit load originally specified. Today you get half and quarter load devices. Thus to see how many devices you can install you simply get the data sheets and add the loads. Look for "UL" on the data sheet. It stands for Unit Load.

Chipkin
Automation Systems

What can go wrong with RS485

What can go wrong with RS485

Lets say you adopted all the best practices for installation of the network but you get intermittent or unacceptable performance because of packet loss, noise, collisions ... Then you should consider hiring an expert to resolve your problems because now you are in the 'Art' part of RS485. These are some of the things they will look at.

Reflections.

> Without a scope and expertise you wont know this is a factor. Its easy and cheap to eliminate. Look at the cable spec. Find the nominal impedance. Buy two resistors of the same value. At each end of the trunk install the resistors between the Tx and Rx terminals. If you don't have obvious ends of the trunk (because you created a star) then we recommend re-cabling to form a linear trunk or we wish you luck.
>
> Some devices have terminating resistors built into them. If the vendor did a poor job, the default is to have the resistor active and they must be disabled unless they are the terminating devices on the network. Read vendor docs.

Biasing , Idle State Biasing, Fail Safe Biasing, Anti Aliasing

> There are a whole string of terms uses as synonyms to describe this phenomenon.
>
> To use two wires (as opposed to full duplex 4 wire) for RS485 each devices transmitter and receiver must be set to an idle state to release the line for others use. Releasing the line means allowing it to 'float'. It must not be allowed to float at any voltage level so devices have pull up/down resistors to pull the line to an allowable 'floating' voltage. (the floating state is also known as the tri-state.) The load presented by other devices on the network affects this floating so the resistor values may need to be changed depending on the number of devices installed and the values of the pull up/down resistors they are using. (You can imagine how tricky its going to be to resolve this).

Chipkin
Automation Systems

What can go wrong with RS485

If a device floats out of the specified range then to other devices it will look like the floating devices isn't floating at all. The other devices will think that it is transmitting or receiving and thus blocking the line.

The simplest way of knowing if this is a factor - Does the device work properly when it is the only device on the network ?. When you install it in the full network other devices or this device stops working properly. This device and/or the pull up/down resistors of other devices are candidates for investigation.

A number of vendors have a range of pull up/down resistors installed and allow you to change the selection using software or jumpers.

Line Drive On / Off

To use two wires for RS485 each devices transmitter and receiver must be set to an idle state to release the line for others use. When a device wants to send it must grab the line. When it has finished sending it must release the line. You can see there are potential problems here. What happens if one device waits too long after sending its last bit before releasing the line - its possible that the other devices will miss some bits of data.

Useful Tip

Other than the addition of terminating resistors to cancel reflections you probably need an expert to help resolve these difficulties. That is why its best to adopt good installation practices.

Chipkin
Automation Systems

MSTP Trunk Topology

Take care of your Trunk Topology

Take care with the topology. The best topology is a single trunk that in-outs on the terminal blocks of each device it connects. What do mean by best ? We mean the choice which is least likely to cause problems.

Best Topology

Simple Multidrop, Unwind the twists as little as possible.(Showing TX conductor for reference only)

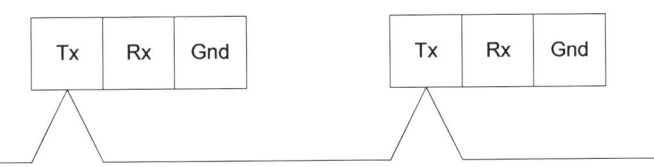

| Tx | Rx | Gnd |

| Tx | Rx | Gnd |

Worse

Making the connections to the RS485 terminals, drops instead of connections starts to give the electrical signals all kinds of complicated paths for reflections and harmonics. Its obvious that if the drops are long and are not twisted then you also have more chance to induce noise. (Showing TX conductor for reference only)

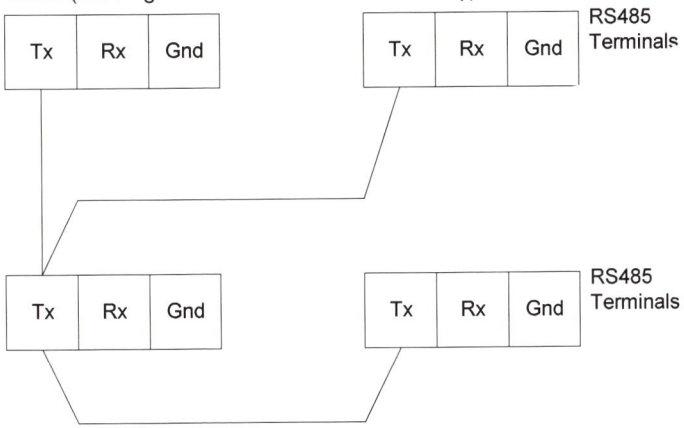

| Tx | Rx | Gnd |

| Tx | Rx | Gnd | RS485 Terminals

| Tx | Rx | Gnd |

| Tx | Rx | Gnd | RS485 Terminals

What can go wrong with RS485

Worst

☞ Avoid Star configurations. They are so much harder to debug when it gets tricky. (Showing TX conductor for reference only)

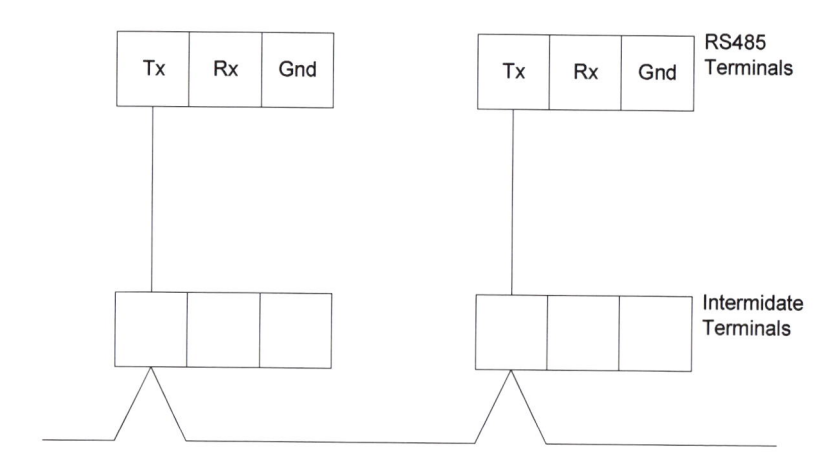

Tx	Rx	Gnd

Tx	Rx	Gnd	RS485 Terminals

Intermidate Terminals

Chipkin
Automation Systems

How MSTP devices are discovered

BACnet MSTP is a token passing protocol. Only nodes with the token are allowed to initiate service requests such as requests for data. A device that receives a request, a request that requires a response, may respond without having the token.

Based on this behavior it is easy to understand the difference between a MSTP master and slave. A slave is a device that can only send responses. A master is an initiator of a service request.

Only allowing masters to initiate a message exchange when they have the token provides a mechanism whereby there can be multiple masters on a network and contention or collision can be avoided. Ethernet uses a different system - it allows collisions and provides a recovery mechanism. Imposing rules on the token passing such as specifying how much a master can do while it has the token provides a mechanism to balance the performance of various devices on a single network.

Back to the question of how new devices are added to the network. If you add a new slave device then you will need to program at least one master on the network to exchange data with that slave.

If you add another master it needs to receive the token before it can act like a master but the other devices on the network need to discover the new device first. Every master on the network, has the job, of periodically polling for a new master.

Each master knows who the next master on a network is, because that is who it will pass the token to. So, each master, must poll for masters that could exist in the address range that exists between its own address and the next master's address. Thus a master addressed as 1 must look for master's in the range 2 to 10 if the next known master is 11. Master number 11 must look for new masters starting at 12 etc. The master with the highest number must try masters ...,127,0...... When a device receives a poll asking if it is a master (Called a 'Poll for Master' message) it replies immediately. In the above example, if master number 1 cant find a master number 2 it should try number 3.

When should it try ? That's a choice that is left up to the implementer of the BACnet protocol on that device. The spec only demands a minimum of 1 Poll for Master every 50 times a master receives or uses the token. The new master

MSTP Discovery cont'd

You can see, that if every master polls for a large number of new masters and they do this often then lots of bandwidth is lost. For this reason BACnet MSTP has a parameter called Max Master. Each master has its own setting for this variable. Typically it is set at 127 but imagine that master number 50 is the highest master on the network, and its Max Master is set to 64. Then it will never discover a new master whose address is larger than its max master, ie it will never discover master with address 65 to 127. This is a common reason why a new device on a network is not discovered.

How often should a master search ? http://www.chipkin.com/articles/how-often-should-a-bacnet-mstp-device-search-for-a-new-master

Passive Slaves
The BACnet Master Slave Token Passing (MSTP) LAN works on a token passing principle. A master node has to gain access to a token to be able to use the transport medium. Only master nodes are allowed to send and receive tokens on the MSTP network. Passive slave nodes on the other hand may only transmit a data frames on the network in response to a request from a master nodes. Passing the token represents overhead in the sense that the messages used for managing the token do not carry data that is useful to automation or monitoring.

On the BACnet MSTP network, frame types are used as a mechanism to allow passive slaves node's, that never holds the token, a means to return replies.

Tip—Mac Addresses

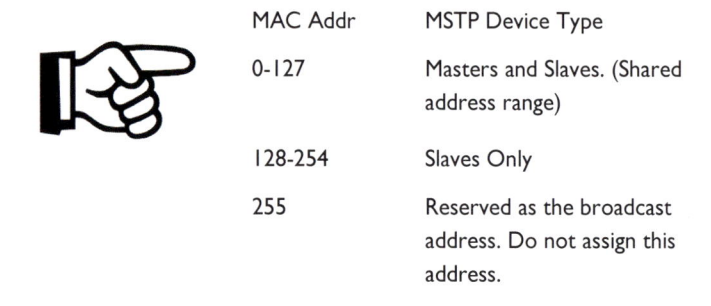

MAC Addr	MSTP Device Type
0-127	Masters and Slaves. (Shared address range)
128-254	Slaves Only
255	Reserved as the broadcast address. Do not assign this address.

Chipkin
Automation Systems

MSTP Slaves vs. Masters

Many vendors do not provide a choice between Master and Slave configuration of their field devices. They are masters and there is nothing you can do about it.

Slaves cannot discover new devices and hand them the token.
"A slave node shall neither transmit nor receive segmented messages."
Slaves can be discovered
Slaves conserve bandwidth.
Masters burn bandwidth

So you must weigh this small one time engineering cost against the huge bandwidth burden using many masters on a network imposes. The charts below give you an idea of how big the burden is.

Based on traffic on a simple network with a single master and a single field device. In test 1 we configured the field device as a Master. In test 2 we configured the field device as a slave.

The frames and frame types that will be ignored by a passive slave node are as follows:

- Frames with a destination address not equal to this station address (TS)
- Frame with destination address equal to 255 (broadcast address) and frame type equal to BACnet_Data_Expecting_Reply, Test_Request or a proprietary type known to the node that expects a reply. (such frames may not be broadcasted)
- Frame types Token, Poll_For_Master, Reply_To_Poll_For_Master, Reply_Postponed or a standard or proprietary frame type not known to this station.

The frames and frame types excepted by a passive slave node are as follows:

- Frames with destination address equal to this station address (TS) and with frame type Receive_data_no_reply, Test_Response or proprietary type known to the node that does not expect a reply.
- Frames with destination address equal to this station address (TS) and with frame type Receive_Data_Expecting_Reply, Test_Request or proprietary type known to this station that expects a reply.

MSTP Slaves vs. Masters

Chart 1 : Typical ratio of payload (APDU) to overhead (Token) when using **Masters**.

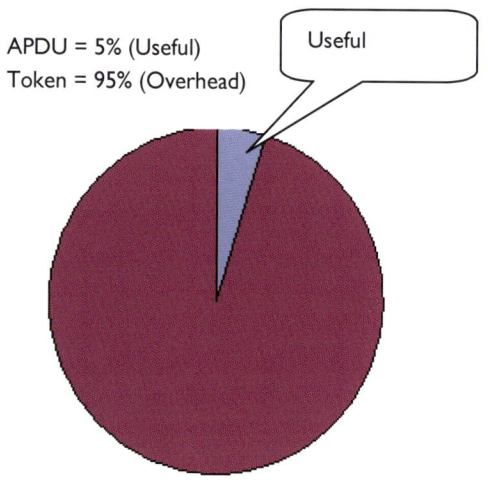

APDU = 5% (Useful)
Token = 95% (Overhead)

Chart 2 : Typical ratio of payload (APDU) to overhead (Token) when using a **Slave** device.

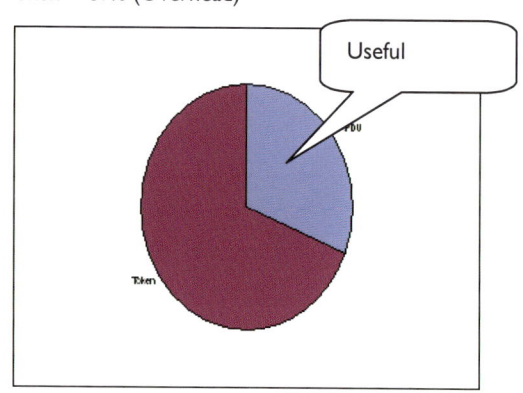

APDU = 31% (Useful)
Token = 69% (Overhead)

Changing the Present Value

All devices on a BACnet network are effectively peers. This means that any device (we take device here to mean any Bacnet capable entity - device or software application) can write to the writable properties of another device's objects. This can result in conflicting commands.

BACnet has a mechanism to resolve the conflict. It differentiates between writable and commandable properties and the conflict resolution only applies to commandable properties. For writable (and non-commandable properties) the last write wins and overwrites any previous writes - there is no conflict resolution.

Useful Tip

Its possible that when you change the set point it will have no effect. The reason is that you may not have sent the new set point with a high enough priority so its using the set point someone else sent instead.

Which properties are commandable and how does the command resolution work ?

- The Present Value of AO, BO, MO objects are always commandable.

- The Present Value of AV, BV, MV objects are commandable if the vendor implemented them that way. It's a vendor choice. You can tell what choice they made by looking for the Priority_Array and Relinquish_Default properties on the object. That's a clue but not a guarantee (we have found). Last resort is their documentation. (good luck).

- A vendor may choose to make any vendor (proprietary) property commandable. If a property is commandable it is required to have appropriately named Priority_Array and Relinquish_Default properties.

Chipkin Automation Systems

Relinquish Default—Worked Example

The Relinquish Default Value is set by the device. The Vendor may choose to make it a writable property in which case in can be changed remotely. Even though the present value is commanded the device stores the commanded value in the priority array and uses the highest priority array slot to set the Present Value.

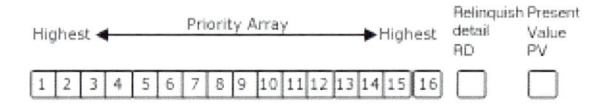

In our example, the device boots, the Priority array slots are all Null (Unused) and this vendor has set the Relinquish Default to 50. Since all the slots are null the device sets the Present Value to the Relinquish Default Value. The Present Value changes to 50.

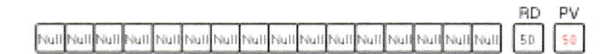

Now a command is sent to set this objects Present Value to 45 at Priority 5. The device sets slot 5 in the Priority Array to 45. It then starts at the highest priority (1) and looks for the 1st non Null slot. It finds slot 5 filled with 45 and sets the Present Value to 45.

Chipkin
Automation Systems

Relinquish Default—Worked Example

Now a new command is sent to set this objects Present Value to 70 at Priority 8. The device sets slot 8 in the Priority Array to 70. It then starts at the highest priority (1) and looks for the 1st non Null slot. It finds slot 5 filled with 45. Thus there is no change to the Present Value to 45.

Null	Null	Null	Null	45	Null	Null	70	Null	Null	Null	Null	Null	Null	Null	Null		RD	PV
																	50	45

Now a command is sent to Relinquish the command at Priority 5. One would hope that the device that sent the original command sent the relinquish command but that is up to you and how you configured you system. When the relinquish command is received, the device sets the corresponding slot in the Priority Array to Null. The device then starts at the highest priority (1) and looks for the 1st non Null slot. The device finds slot 8 filled with 70. It changes the Present Value to 70.

Null	Null	Null	Null	Null	Null	Null	70	Null	Null	Null	Null	Null	Null	Null	Null		RD	PV
																	50	70

The most recent command at a specific priority wins. Here a command is sent to set the Present Value to 80 at priority 8. The device overrides slot 8 in the array with the new value. In this case it is also the highest priority slot that is used so the device updates the Present Value to 80.

Null	Null	Null	Null	Null	Null	Null	80	Null	Null	Null	Null	Null	Null	Null	Null		RD	PV
																	50	80

Relinquish Default—Worked Example

Finally, a command is sent to relinquish the command at priority 8. Slot 8 is set to Null and when the device looks through the priorty array it finds it all empty and it thus uses the Relinquish Default value to set the Present Value to 50.

Null	Null	Null	Null	Null	Null	Null	Null	Null	Null	Null	Null	Null	Null	Null	Null	RD	PV
																50	50

This mechanism can be complicated if the object has minimum on/off times.

Chipkin
Automation Systems

Trouble Shooting BACnet IP / Ethernet

Required tools:

    Hub or Supervised Switch
    Wireshark – Free Download
       http://www.wireshark.org/download.html

Warning !

You might not capture the traffic if you don't use a hub. Read the article on hub and switches to understand why. http://www.chipkin.com/articles/hubs-vs-switches-using-wireshark-to-sniff-network-packets

Useful Tip

You can select the packets you capture to reduce log file size by defining a capture filter before you start the capture. We suggest you avoid this. If you are short of space you can select which packets you save.

Useful Tip

You can select which packets you view from the total log by defining a display filter. You can select which packets to save in the log files.

Useful Tip

You can search for particular packets.

How to Capture with Wireshark

Step 1: Capture – Main Menu

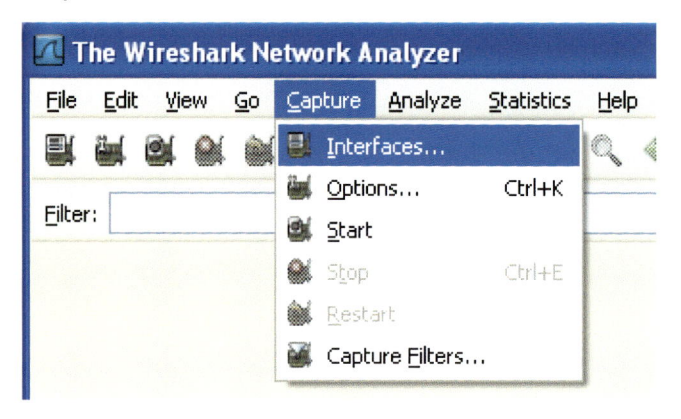

Step 2: Interfaces - On Capture Menu

    a. You get a list of network adapters. Pick the one connected to the network of interest. Its probably not the wireless adapter. Most often it's the adapter with the packet count increasing

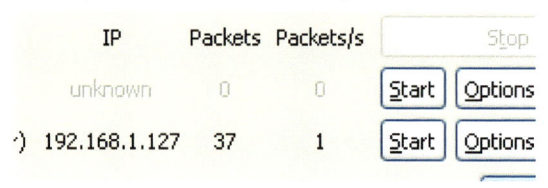

b. Select the Start button or
c. Select the options button to define a capture filter. Define the filter and click start.

How to Capture with Wireshark

Step 3: A list of packets accumulates on the screen

	Time	Source	Destination	Protocol	Info
74	31.910539	192.168.1.127	79.42.90.233	UDP	Source
75	32.234648	79.42.90.233	192.168.1.127	UDP	Source
76	33.150373	Cisco-Li_3a:27:b8	Broadcast	ARP	who has
77	34.145611	Cisco-Li_3a:27:b8	Broadcast	ARP	who has
78	34.154012	192.168.1.127	192.168.1.81	BACnet-A	Confirm
79	34.167543	192.168.1.81	192.168.1.127	BACnet-A	Complex
80	35.146249	Cisco-Li_3a:27:b8	Broadcast	ARP	who has
81	36.177646	Cisco-Li_3a:27:b8	Broadcast	ARP	who has
82	37.177569	Cisco-Li_3a:27:b8	Broadcast	ARP	who has
83	38.178137	Cisco-Li_3a:27:b8	Broadcast	ARP	who has
84	39.647738	Cisco-Li_3a:27:b8	Broadcast	ARP	who has
85	40.639518	Cisco-Li_3a:27:b8	Broadcast	ARP	who has
86	41.444944	192.168.1.127	192.168.1.81	BACnet-A	Confirm
87	41.457484	192.168.1.81	192.168.1.127	BACnet-A	SimpleA
88	41.640101	Cisco-Li_3a:27:b8	Broadcast	ARP	who has

Step 4: Apply a Display Filter. More on display filters later. For now simply type bacnet into the filter field and click apply. .

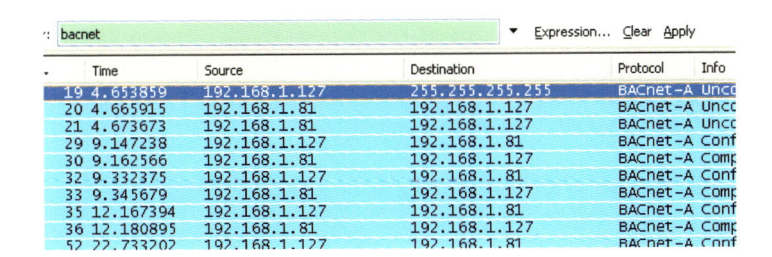

Filter: bacnet   Expression... Clear Apply

	Time	Source	Destination	Protocol	Info
19	4.653859	192.168.1.127	255.255.255.255	BACnet-A	Unco
20	4.665915	192.168.1.81	192.168.1.127	BACnet-A	Unco
21	4.673673	192.168.1.81	192.168.1.127	BACnet-A	Unco
29	9.147238	192.168.1.127	192.168.1.81	BACnet-A	Conf
30	9.162566	192.168.1.81	192.168.1.127	BACnet-A	Comp
32	9.332375	192.168.1.127	192.168.1.81	BACnet-A	Conf
33	9.345679	192.168.1.81	192.168.1.127	BACnet-A	Comp
35	12.167394	192.168.1.127	192.168.1.81	BACnet-A	Conf
36	12.180895	192.168.1.81	192.168.1.127	BACnet-A	Comp
52	22.733202	192.168.1.127	192.168.1.81	BACnet-A	Conf

Chipkin
Automation Systems

How to Capture with Wireshark

Step 5: Find the packet you are interested in. Click on it to select it. A breakout of the selected packet's data is shown below the packet list.

```
 52 22.733202    192.168.1.127    192.168.1.81
 53 22.746262    192.168.1.81     192.168.1.127
 61 27.035724    192.168.1.127    192.168.1.81
 62 27.050641    192.168.1.81     192.168.1.127
 78 34.154012    192.168.1.127    192.168.1.81
 79 34.167543    192.168.1.81     192.168.1.127
 86 41.444944    192.168.1.127    192.168.1.81
 87 41.457484    192.168.1.81     192.168.1.127
```

```
⊞ Frame 52 (73 bytes on wire, 73 bytes captured)
⊞ Ethernet II, Src: AsustekC_b7:84:06 (00:23:54:b7:84:0
⊞ Internet Protocol, Src: 192.168.1.127 (192.168.1.127)
⊞ User Datagram Protocol, Src Port: bacnet (47808), Dst
⊞ BACnet Virtual Link Control
⊞ Building Automation and Control Network NPDU
⊟ Building Automation and Control Network APDU
     0000 .... = APDU Type: Confirmed-Request  (0)
  ⊞ .... 0000 = PDU Flags: 0x00
     .000 .... = Max Response Segments accepted: Unspec
```

Step 6: You can break out the level of detail by expanding the sections of the packet.

> Think of a bacnet packet as a letter you send to a bacnet device. When you take it to the bacnet post office. The clerk says he does not understand the address. He passes it to the UDP clerk. The UDP clerk takes your letter and puts it in a bigger envelope. He addresses the envelope with a UDP address. He passes it to the IP post office clerk. The IP clerk takes your letter and puts it in a bigger envelope. He addresses the envelope with an IP address. He passes it to the Ethernet post office clerk....

Chipkin
Automation Systems

How to Capture with Wireshark

Step 6: continued

The Ethernet clerk takes your letter and puts it in a bigger envelope. He addresses the envelope with a hardware address and sends it to that computer. When it arrives the process is reversed until finally the contents are passed to the bacnet application.

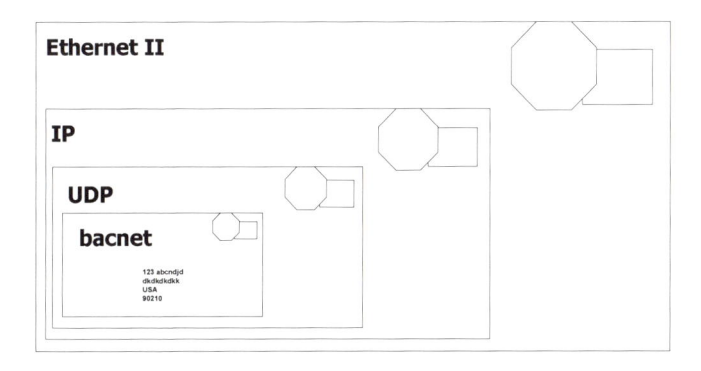

```
Ethernet II

    IP

        UDP

            bacnet
                123 abcndjd
                dkdkdkdkk
                USA
                90210
```

Ethernet packets contain packets from other higher level protocols nested inside each other. You drill down to see the detail you want.

In the example below you can see the bacnet packet nested inside a UDP (User Datagram Protocol) which is nested inside an IP protocol packet which is in turn nested inside an Ethernet packet. Drilling down into the Bacnet packet you can see the concept is carried even further

Chipkin
Automation Systems

How to Capture with Wireshark

The bacnet service is a write to analog input #1. That is wrapped up with some information about the device – the device number and network number is contained inside the NPDU.

```
⊞ Frame 52 (73 bytes on wire, 73 bytes captured)
⊞ Ethernet II, Src: AsustekC_b7:84:06 (00:23:54:b7:84:0
⊞ Internet Protocol, Src: 192.168.1.127 (192.168.1.127)
⊞ User Datagram Protocol, Src Port: bacnet (47808), Dst
⊞ BACnet Virtual Link Control
⊞ Building Automation and Control Network NPDU
⊟ Building Automation and Control Network APDU
     0000 .... = APDU Type: Confirmed-Request (0)
  ⊞ .... 0000 = PDU Flags: 0x00
     .000 .... = Max Response Segments accepted: Unspeci
     .... 0101 = Size of Maximum ADPU accepted: Up to 14
     Invoke ID: 4
     Service Choice: writeProperty (15)
  ⊞ ObjectIdentifier: analog-input object, 1
  ⊞ Property Identifier: present-value
  ⊟ propertyValue
     ⊞ Opening Tag: 3
     ⊞ present-value: 20.000000 (Real)
     ⊞ Closing Tag: 3
  ⊟ Priority: (Unsigned) 8
     ⊞ Context Tag: 4, Length/Value/Type: 1
```

How to Capture with Wireshark

Step 7: Drill down to see the BACnet info

```
- Building Automation and Control Network NPDU
    Version: 0x01 (ASHRAE 135-1995)
  + Control: 0x24
    Destination Network Address: 99         ← bacnet network #
    Destination MAC Layer Address Length: 1
    DADR: 16  ← Device Address
    Hop Count: 255
- Building Automation and Control Network APDU
    0000 .... = APDU Type: Confirmed-Request  (0)
  + .... 0000 = PDU Flags: 0x00
    .000 .... = Max Response Segments accepted: Unspecified (0)
    .... 0101 = Size of Maximum ADPU accepted: Up to 1476 octets (fits in an ISO 8802
    Invoke ID: 4
    Service Choice: writeProperty (15)       ← Service, object, property and value
  + ObjectIdentifier: analog-input object, 1
  + Property Identifier: present-value
  - propertyValue
    + Opening Tag: 3
    + present-value: 20.000000 (Real)
```

How to Filter with Wireshark

Before you start a capture you can specify a capture filter. The effect of the filter is to prevent all packets being captured. Doing this can save space when you save the log and it might make it easier to find the packets you are interested in. However, there is some risk that you might filter out the packets of interest.

For example, a BACnet device might not operate correctly because it is being hammered with packets from another protocol being sent incorrectly to the BACnet device. Our advise is to capture as much as possible and then filter what is displayed.

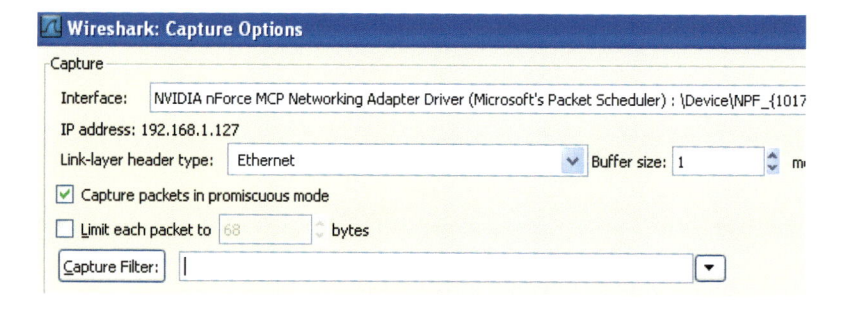

Chipkin Automation Systems

How to Filter with Wireshark cont'd

Here are some sample filters
Examples

    Capture only traffic to or from IP address 172.18.5.4:

    host 172.18.5.4

    Capture only traffic to or from IP address 172.18.5.4 but exclude all FieldServer RUINET messages

    host 192.168.1.81 and port not 1024

    Capture traffic to or from a range of IP addresses:

    net 192.168.0.0/24

    or

    net 192.168.0.0 mask 255.255.255.0

    Capture traffic from a range of IP addresses:

    src net 192.168.0.0/24

    or

    src net 192.168.0.0 mask 255.255.255.0

How to Filter with Wireshark cont'd

Capture traffic from a range of IP addresses:

    src net 192.168.0.0/24

    or

    src net 192.168.0.0 mask 255.255.255.0

Capture traffic to a range of IP addresses:

    dst net 192.168.0.0/24

    or

    dst net 192.168.0.0 mask 255.255.255.0

Capture only bacnet traffic: Assumes every device is compliant and is using the standard port.

    port 47808

Finding Packets in Wireshark

Useful Tip

Its easy to sort packets by source or destination IP, Click the column headings.

Useful Tip

You can mark, packets you find interesting. Then later you can save, display or print the marked packets.

Searching :

The problem is that you can only specify one text string and hence you can only specify one search criteria. For example in the search below, all packets that contain the string = 'analog object (1)' will be found irrespective of the device, ip address ..etc.

Some Useful Search Strings

writeProperty

objectIdentifier: analog-input object, 1
objectIdentifier: analog-output object, 11
objectIdentifier: binary-input object, 10041
objectIdentifier: binary-output object, 45

Finding Packets in Wireshark cont'd

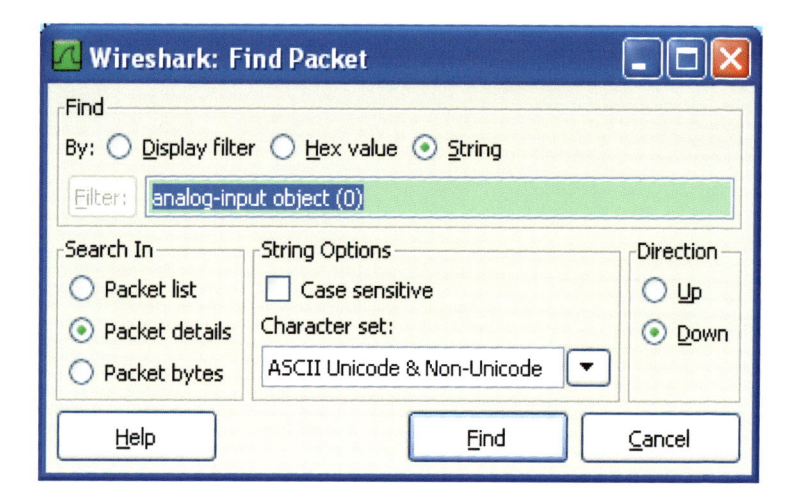

Wireshark - Display Filtering

Useful Tip

Any capture filter can be used as a display filter.

Chipkin
Automation Systems

Find packets using Display Filters

Wireshark - Display Filtering

Useful Tip

Any capture filter can be used as a display filter.

Looking for failures:
Try this search string. Type5 are errors, Type6 are Reject messages and type 7 are abort messages.

bacapp.type == 5 || bacapp.type == 6 || bacapp.type == 7

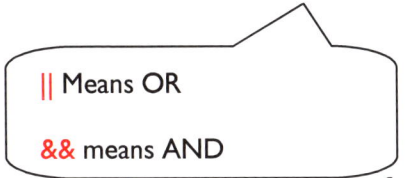

> || Means OR
>
> && means AND

Looking for messages which specify particular objects types:
Try these search strings. 0=AI, 1=AO, 2=AV, 3=BI, 4=BO, 5=BV, 6=Calendar Object, 7=Command Object, 8=Device Object, 9=Event Enrollment, 10=File, 11=Group, 12=Loop, 13=MI, 14=MO,17=Schedule,19=MV, 20=Trend Log

bacapp.objectType == 0

Find packets using Display Filters

Looking for a particular Object: In this example all messages which reference AI(1) are listed.

bacapp.instance_number == 1 && bacapp.objectType == 0

Looking for messages to/from particular devices

ip.addr == 192.168.1.90 (Sent to/Sent From)
ip.dst_host == "192.168.1.90"
ip.src_host == "192.168.1.90"

You can use the expression builder to build filter expressions

Expression Builder

Chipkin
Automation Systems

Find packets using Display Filters

From the drop down list of protocols there are two specifically related to **BACnet**. They are shown below

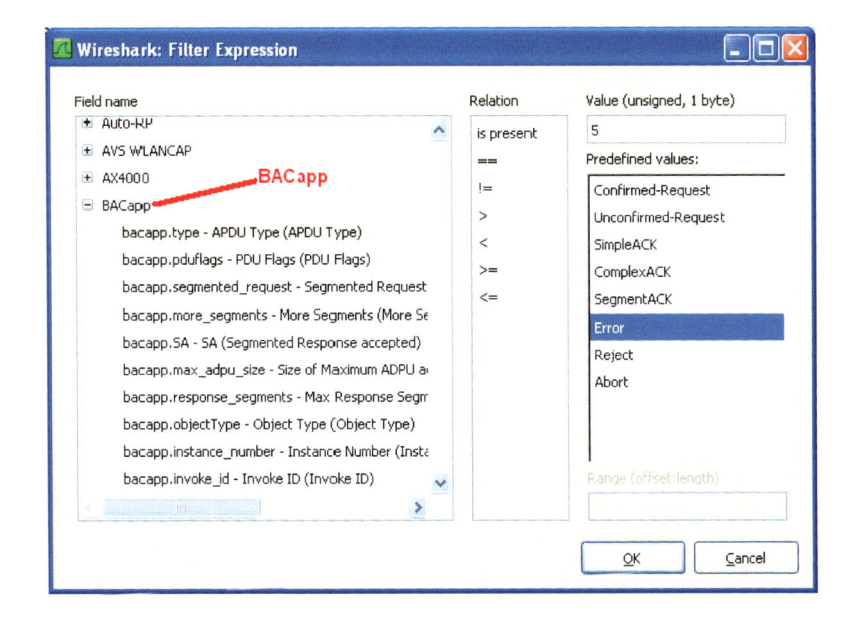

Find packets using Display Filters

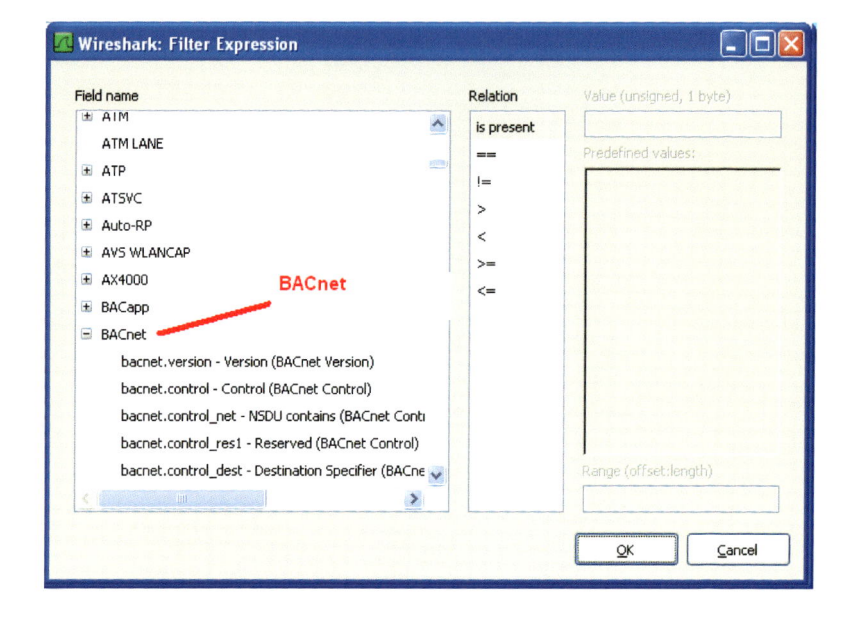

Trouble Shooting BACnet MSTP

Trouble Shooting BACnet MSTP

One badly behaved device on a MSTP trunk can cause a collapse in performance.

To understand why consider this one example of what can go wrong. RS485 is s shared trunk. When two devices transmit at the same time this causes a collision. In simple terms think of the messages interfering with each other and corrupting each other so neither message is recognizable. To prevent this happening and to allow for multiple master on a network, BACnet has chosen a token based system. Only devices with the token can initiate a message transaction. To keep things running smoothly, BACnet has some demanding timing requirements. For example, a device passes the token on. The receiving device has 15msec to use the token. Lets say it responds in 18msec. During that interval, the original device doesn't see the token being used so it send the token again. As it sends it, the 2nd device uses the token (3 msec late). Now the messages from the 1st and 2nd device clash. The 2nd device thinks it has the token and starts to use it. The 1st device thinks the token got lost and starts to poll for a new master. Messages clash, causing contention and eventually both devices recover to a valid state using the protocol rules. These rules require waiting various timeout periods. All the waiting, collisions and recovery waste time and bandwidth. If this happens often (as it will because one device doesn't meet the spec) then you can lose significant bandwidth.

Trouble Shooting BACnet MSTP

Strategy

    Capture Traffic using USB-485 converter
        Use a tool like BACspy
        Or
        Use Hyperterminal

    Analyze Traffic
        Use a tool like BACspy online or Offline
        Or
        Perform a manual analysis

Chipkin
Automation Systems

Trouble Shooting BACnet MSTP

Step 1: Configure HyperTerminal

Select the COM port which corresponds to your USB converter
Set the Baud Rate correctly
Set the other parameters as shown

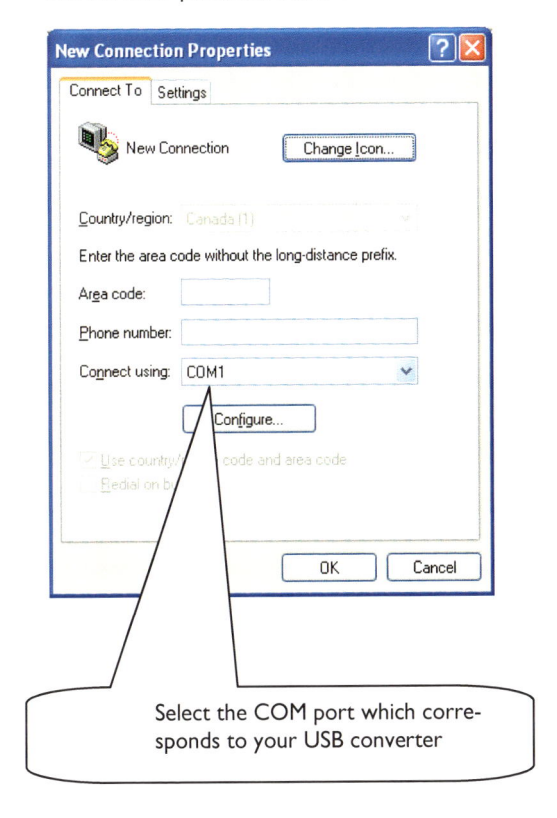

Select the COM port which corresponds to your USB converter

Chipkin
Automation Systems

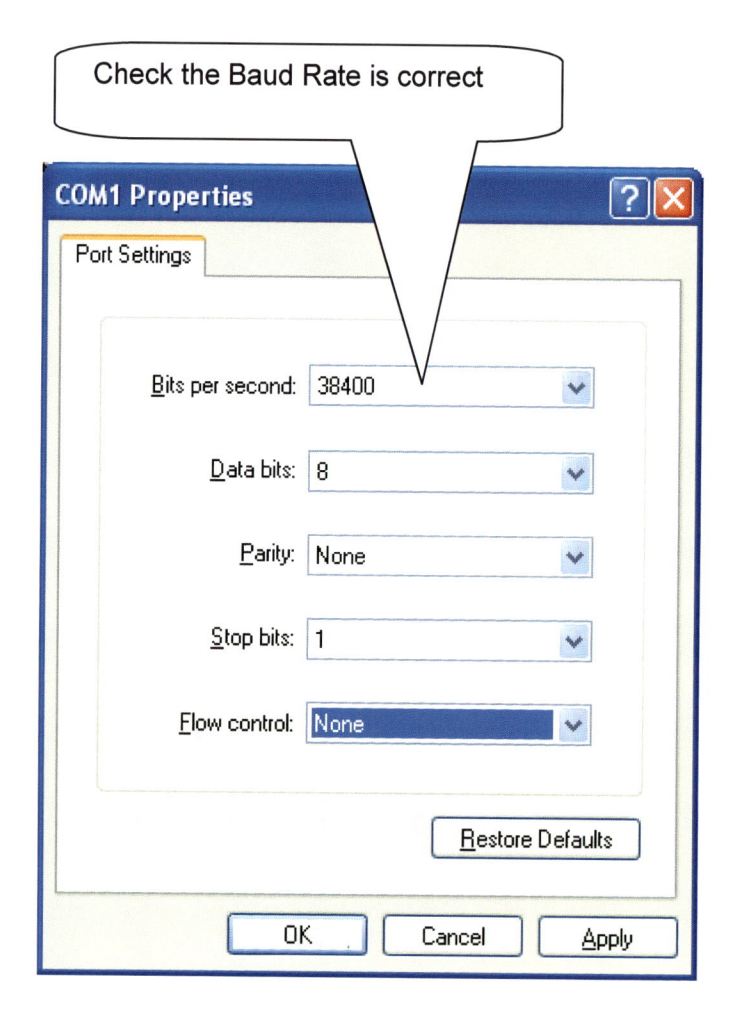

Trouble Shooting BACnet MSTP

Check the Baud Rate is correct

COM1 Properties

Port Settings

- Bits per second: 38400
- Data bits: 8
- Parity: None
- Stop bits: 1
- Flow control: None

Restore Defaults

OK   Cancel   Apply

Trouble Shooting BACnet MSTP

Step 2: Capture to File

Use this menu to start and stop the capture

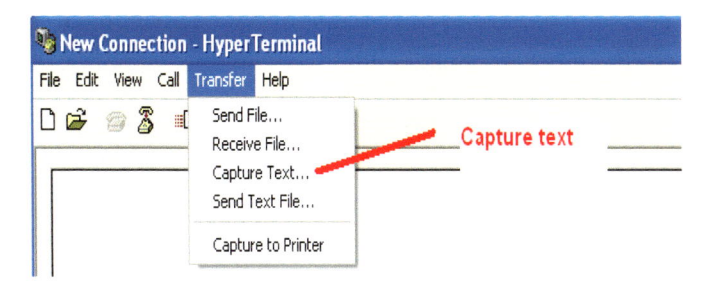

As the messages are captured you will see these non-human readable characters fill the screen.

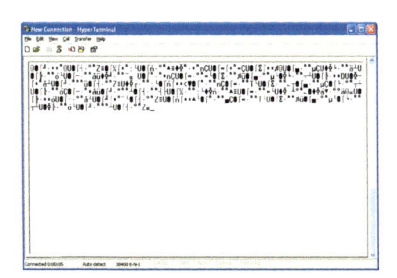

Use this menu to start and stop the capture

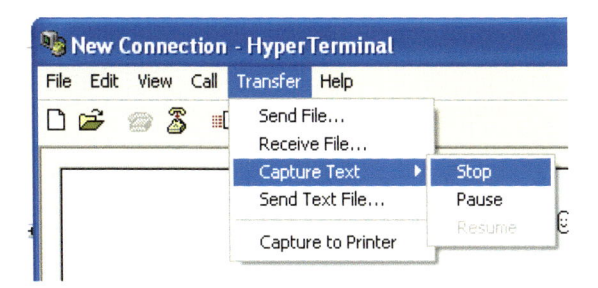

Trouble Shooting BACnet MSTP

Step 3: Inspect the captured data before you leave site to see if it usable.

You need a viewer capable of displaying the hex bytes. Here is a free download and hex viewer. Open the log file.

To ensure you captured useful data look for messages that begin 55 FF. All BACnet MSTP messages begin with the same codes. If you don't see any then this may mean you captured at the wrong baud rate or some other setting is wrong. It is also possible that there is no BACnet communication.

```
HexDump32 - Binary Content Viewer
File  About
00000000   03 02 00 00 00 14 79 E7-03 01 00 00 00 14 3D E7   ......y........=.
00000010   55 FF 00 0B 01 00 00 BE-55 FF 05 01 0B 00 0D 76   U.......U......v
00000020   01 04 00 03 55 0C 0C 01-00 00 01 19 55 18 43 43   ....U.......U.CC
00000030   55 FF 06 0B 01 00 10 5E-01 00 30 55 0C 0C 01 00   U......^..0U....
00000040   00 01 19 55 3E 91 00 3F-4A EA 55 FF 00 01 0B 00   ...U>..?J.U.....
00000050   00 14 55 FF 01 05 01 00-00 4F 55 FF 00 0B 01 00   ..U......OU.....
00000060   00 BE 55 FF 00 01 0B 00-00 14 55 FF 01 06 01 00   ..U.......U.....
00000070   00 D7 55 FF 00 0B 01 00-00 BE 55 FF 00 01 0B 00   ..U.......U.....
00000080   00 14 55 FF 01 07 01 00-00 5E 55 FF 00 0B 01 00   ..U......^U.....
00000090   00 BE 55  Displayed as Hex       'F 01 08 01 00   ..U.......U.....
000000A0   00 A0 55                          'F 05 01 0B 00   ..U.......U.....
000000B0   0D 76 01 04 00 03 55 0C-0C 01 00 00 02 19 55 7B   .v....V.......U{
000000C0   7A 55 FF 06 0B 01 00 10-5E 01 00 30 56 0C 0C 01   zU......^..0V...
000000D0   00 00 02 19 55 3E 91 00-3F D3 4C 55 FF 00 01 0B   ....U>..?.LU....
000000E0   00 00 14 55 FF 01 09 01-00 00 29 55 FF 00 0B 01   ...U......)U....
000000F0   00 00 BE 55 FF 00 01 0B-00 00 14 55 FF 01 0A 01   ...U.......U....
00000100   00 00 B1 55 FF 00 0B 01-00 00 BE 55 FF 00 01 0B   ...U.......U....
00000110   00 00 14 55 FF 01 00 01-00 00 BE 55 FF 00 01 0B   ...U.......U....
00000120   00 00 14 55 FF 00 0B 01-00 00 BE 55 FF 00 01 0B   ...U.......U....
00000130   00 00 14 55 FF 00 0B 01-00 00 BE 55 FF 00 01 0B   ...U.......U....
00000140   00 00 14 55 FF 00 0B 01-00 00 BE 55 FF 00 01 0B   ...U.......U....
```

http://hexdump32.salty-brine-software.qarchive.org/_download2.html

Trouble Shooting BACnet MSTP

Step 4: You can use CAS BACspy

BACspy automates this process of capture and analysis.

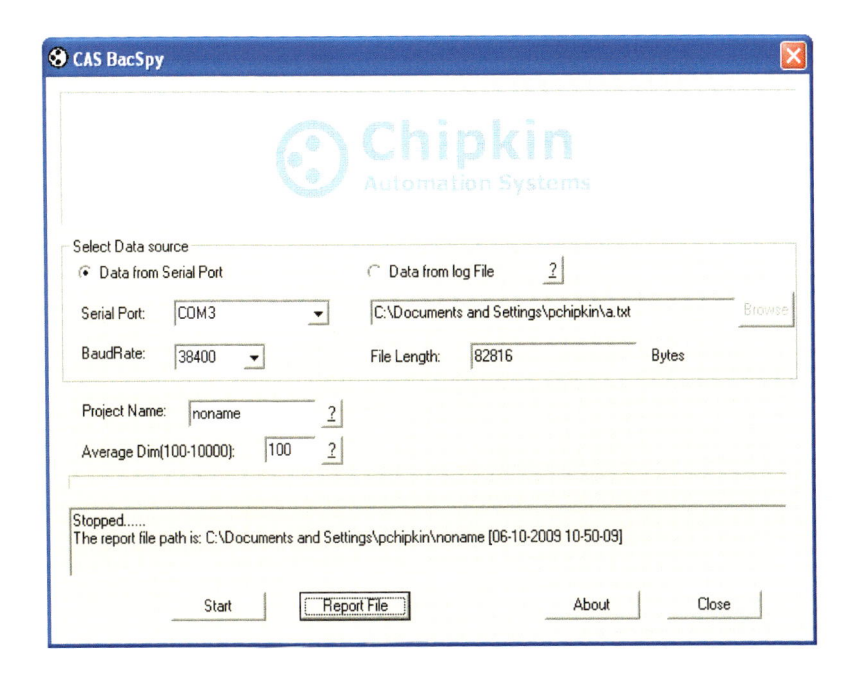

BACspy can also be used to analyze data captured using HyperTerminal

Chipkin Automation Systems

Trouble Shooting BACnet MSTP

> These longer messages are application layer messages such as Read the present value of AI(2) on Device 37001.
>
> To see a breakout of this application layer data you need the enhanced version of BACspy which is included in the CAS BACnet Explorer.

Created by: BACSpy, v1.1.0
Created on [06-10-2009 10:50:09]
Chipkin.com

Raw MSTP Frame log file

```
Frame 000000 [06-10-2009 10:50:09:026] S_MAC-1  D_MAC-7   Poll                         /Valid  55 FF 00 0B 01 00 00 BE
Frame 000001 [06-10-2009 10:50:09:088] S_MAC-1  D_MAC-11  Token                        /Valid  55 FF 00 01 0B 00 00 14
Frame 000002 [06-10-2009 10:50:09:094] S_MAC-11 D_MAC-8   Poll For                     /Valid  55 FF 01 09 01 0B 00 00 29
Frame 000003 [06-10-2009 10:50:09:104] S_MAC-1  D_MAC-1   Token                        /Valid  55 FF 00 0B 01 00 00 BE
Frame 000004 [06-10-2009 10:50:09:167] S_MAC-1  D_MAC-11  Token                        /Valid  55 FF 00 01 0B 00 00 14
Frame 000005 [06-10-2009 10:50:09:173] S_MAC-11 D_MAC-9   Poll For Master              /Valid  55 FF 01 0A 01 00 00 B1
Frame 000006 [06-10-2009 10:50:09:183] S_MAC-1  D_MAC-11  Token                        /Valid  55 FF 00 0B 01 00 00 BE
Frame 000007 [06-10-2009 10:50:09:247] S_MAC-1  D_MAC-11  Poll For Master              /Valid  55 FF 05 01 0B 00 0D 76 01 04 00 03 2C 0C 0
Frame 000008 [06-10-2009 10:50:09:253] S_MAC-11 D_MAC-1   Token                        /Valid  55 FF 06 0B 01 00 13 5F 01 00 30 2C 0C 0
Frame 000009 [06-10-2009 10:50:09:263] S_MAC-1  D_MAC-10  Token                        /Valid  55 FF 00 01 0B 00 00 14
Frame 000010 [06-10-2009 10:50:09:327] S_MAC-1  D_MAC-11  Token                        /Valid  55 FF 00 0B 01 00 00 BE
Frame 000011 [06-10-2009 10:50:09:334] S_MAC-11 D_MAC-1   Data Expecting Reply         /Valid  55 FF 00 01 0B 00 00 14
Frame 000012 [06-10-2009 10:50:09:374] S_MAC-1  D_MAC-11  Data Not Expecting Reply     /Valid  55 FF 00 0B 01 00 00 BE
Frame 000013 [06-10-2009 10:50:09:381] S_MAC-11 D_MAC-1   Token                        /Valid  55 FF 00 01 0B 00 00 14
Frame 000014 [06-10-2009 10:50:09:390] S_MAC-1  D_MAC-11  Token                        /Valid  55 FF 00 0B 01 00 00 BE
Frame 000015 [06-10-2009 10:50:09:406] S_MAC-11 D_MAC-1   Token                        /Valid  55 FF 00 01 0B 00 00 14
Frame 000016 [06-10-2009 10:50:09:413] S_MAC-1  D_MAC-11  Token                        /Valid  55 FF 00 0B 01 00 00 BE
Frame 000017 [06-10-2009 10:50:09:422] S_MAC-11 D_MAC-1   Token                        /Valid  55 FF 00 01 0B 00 00 14
Frame 000018 [06-10-2009 10:50:09:429] S_MAC-1  D_MAC-11  Token                        /Valid  55 FF 00 0B 01 00 00 BE
Frame 000019 [06-10-2009 10:50:09:438] S_MAC-11 D_MAC-1   Token                        /Valid  55 FF 00 01 0B 00 00 14
```

Hubs vs Switches

Hubs vs Switches - Using WireShark to sniff network packets

Gotcha #1 : Use a hub not a switch

Why: Switches don't copy all messages to all ports. They try and optimize traffic so when they learn which port a device is connected to they send all messages intended for that device to that port and stop copying to all ports. (The jargon they use for this function is 'learning mode')

How do you know it's a hub: Just because it calls itself a hub doesn't mean it is one.

- If it says full-duplex in the product description it's probably not a hub.
- A switch that allows you to turn off the learning mode is effectively a hub.
- A switch with a monitored port copies all messages to the monitored port and thus you can use that port as if it were a hub.
- If it says 'switch' and you cant turn off learning mode and it doesn't have a monitor port then it is not a hub.
- A router is never a hub.

Hubs vs Switches

Gotcha #2 : Mixing 10 and 100 mbits/sec can cause problems.
Not all hubs copy 10mbit messages to 100mbit ports and vice versa. Use a 10mbit/sec hub if you are on a mixed network – almost all other faster devices are speed sensing and will downgrade themselves to 10mbits/sec and thus you will see all the packets. This is not true of some building automation engines where the speed of the port is configured.

You can work around this problem by connecting higher speed devices to a self sending switch/hub and then connect that switch/hub to the 10mbit hub.

Recommended Hubs

10Mbit/sec Networks - DX-EHB4 - 4 Port 10 Mbps HUB
Netgear - DS104 Dual Speed HUB
10Mbit/sec Networks – D-LINK DE-805TP

Chipkin
Automation Systems

Resistors

Beer and Vodka Can Help You Select a Terminating Resistor

Try this mnemonic if you are trying to remember the resistor color codes:

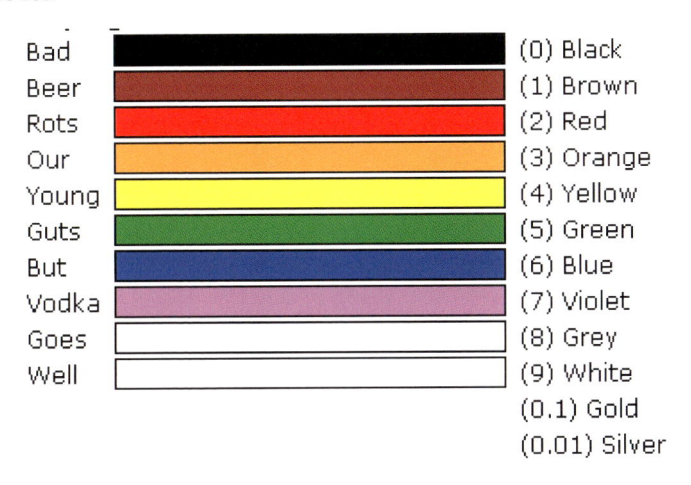

Mnemonic	Color	Value
Bad	Black	(0) Black
Beer	Brown	(1) Brown
Rots	Red	(2) Red
Our	Orange	(3) Orange
Young	Yellow	(4) Yellow
Guts	Green	(5) Green
But	Blue	(6) Blue
Vodka	Violet	(7) Violet
Goes	Grey	(8) Grey
Well	White	(9) White
	Gold	(0.1) Gold
	Silver	(0.01) Silver

Note: If you're missing a tolerance band that implies that the tolerance is 20%.

Which end do you start reading the color bands?
There are usually two ways:

1) If one of the bands at the end of the sequence is further apart then that is the tolerance band - start from the opposite end.
2) If all the bands are closer to one side of the resistor then start from that end - the tolerance band is the last one your read.

Resistors

Find Tollerance Band (Usually Separated) and work from other side

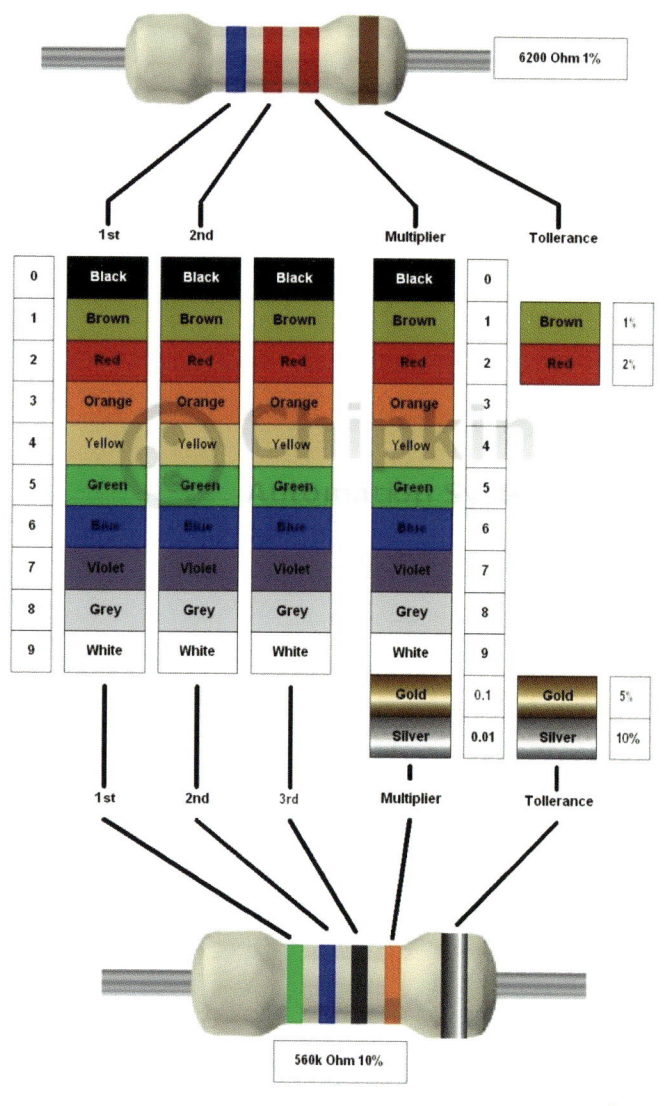

	1st	2nd	Multiplier		Tollerance	
0	Black	Black	Black	0		
1	Brown	Brown	Brown	1	Brown	1%
2	Red	Red	Red	2	Red	2%
3	Orange	Orange	Orange	3		
4	Yellow	Yellow	Yellow	4		
5	Green	Green	Green	5		
6	Blue	Blue	Blue	6		
7	Violet	Violet	Violet	7		
8	Grey	Grey	Grey	8		
9	White	White	White	9		
			Gold	0.1	Gold	5%
			Silver	0.01	Silver	10%

6200 Ohm 1%

560k Ohm 10%

1st — 2nd — 3rd — Multiplier — Tollerance

Page 76

Chipkin Automation Systems

Terminating and Biasing Resistors

What should you carry with you to site? (for communication networks purposes)

For Terminations	
Value	Tolerance
75 Ohm	5%
100 Ohm	5%
120 Ohm	5%

For Biasing
Value
10k Ohm
4k7 Ohm
2k4 Ohm
1k Ohm
560 Ohm
30 Ohm

How to buy resistors

Buy a series - E12 or E24 (they come in packs and provide a comprehensive range of resistors).

Chipkin Automation Systems

Terminating and Biasing Resistors

How to make a resistance value even if you don't have the correct resistor in your toolbox.

Series

$$R_3 = R_1 + R_2$$

- R1 and R2 connected in series could be replaced with one resistor of resistance R3.
- Using 2 or more resistors instead of one allows you to achieve custom resistances.
  Less resistors are required when used in series to achieve greater resistance than when used in parallel

Parallel

- R1 and R2 connected in parallel act as one resistor with a custom resistance of R3.
- This method allows you to get custom resistances using the formula.
- R3 will always be smaller than R1 and R2, so more resistors are required to achieve higher custom resistances.

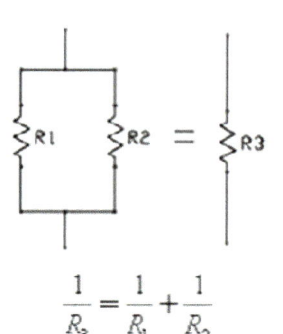

$$\frac{1}{R_3} = \frac{1}{R_1} + \frac{1}{R_2}$$

Chipkin
Automation Systems

CAS BACnet Software

BACspy

MSTP performance and trouble shooting tool. Reports used MAC addresses, who talks to who, who breaks rules .....

BACsql

Provides a myslq or sqlite interface to a BACnet network

BACwatchdog

Monitor who read or writes to a device, object or property.

BACnet Explorer

Discover, browse, monitor, document BACnet networks

Virtual Thermostat / Lighting Controllers

Windows temperature and lighting controllers that talk BACnet.

MSTP / IP / Eth Stack (Protocol Library)

Support for C++ and C#. Tested on Windows and Linux.

Chipkin Automation Systems

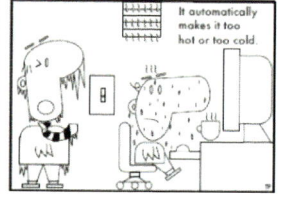

Chipkin Automation Systems

3495 Cambie St, #211
Vancouver, BC
Canada,
V5Z 4R3

Phone: 866-383-1657
E-mail: bacnet@chipkin.com

www.chipkin.com

CPSIA information can be obtained
at www.ICGtesting.com
Printed in the USA
LVIC06n1539030214
372108LV00009BA/58